Cold Air Distribution System Design Guide

This publication was prepared under ASHRAE research project RP-849. Cognizant ASHRAE Technical Committee: TC 6.9, Thermal Storage.

Allan T. Kirkpatrick is a professor of mechanical engineering at Colorado State University, Fort Collins, Colo. He teaches and conducts research in applied thermal sciences. He received his Ph.D. from the Massachusetts Institute of Technology. He is a member of the following ASHRAE Technical Committees: TC 4.10, Indoor Environmental Calculations; TC 5.3, Room Air Diffusion; and TC 6.9, Thermal Storage.

James S. Elleson is principal of Elleson Engineering, Black Earth, Wis. He has been involved in research, design, analysis, and performance evaluation of thermal storage and cold air distribution systems since 1986. He is a principal author of several design guides, ASHRAE papers, and EPRI research reports on these topics.

COLD AIR DISTRIBUTION

SYSTEM DESIGN GUIDE

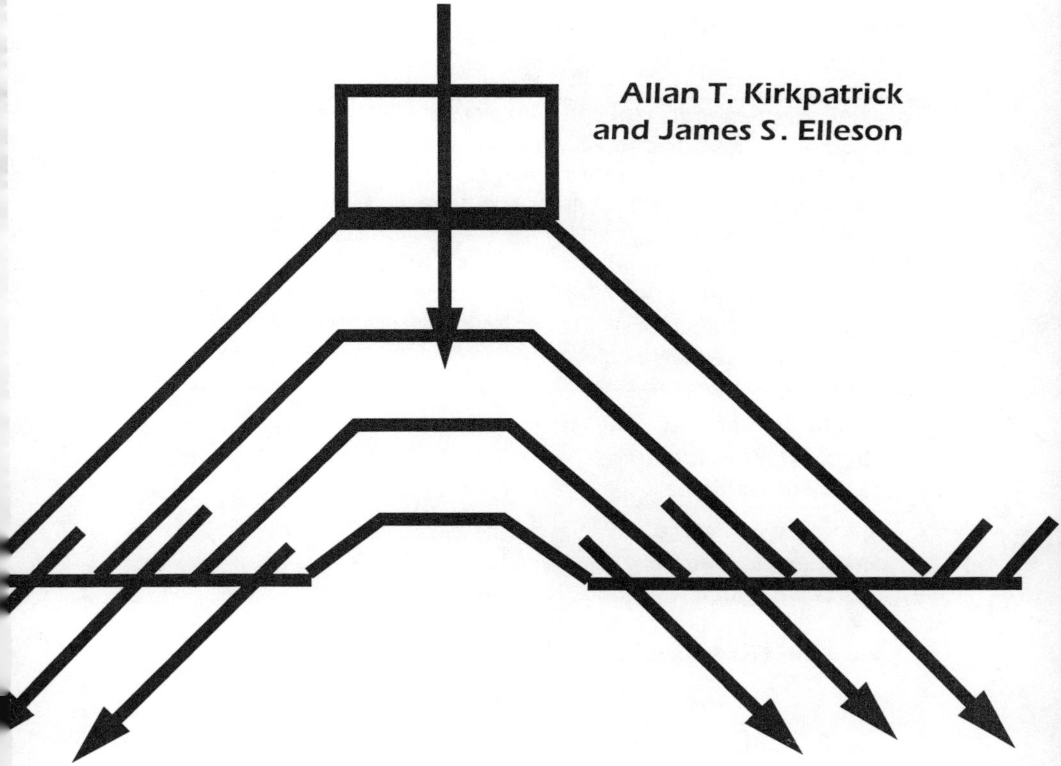

Allan T. Kirkpatrick
and James S. Elleson

American Society of Heating, Refrigerating
and Air-Conditioning Engineers, Inc.

ISBN 1-883413-37-0

1791 Tullie Circle N.E.
Atlanta, Georgia 30329

Printed in the United States of America

ASHRAE STAFF

SPECIAL PUBLICATIONS

Mildred Geshwiler
Editor

Michelle Moran
Assistant Editor

Stefanie Frick
Secretary

Jeri Alger
Managing Editor

Malikah Berry
Editorial Assistant

PUBLISHING SERVICES

Scott Zeh
Manager

Frank M. Coda
Publisher

W. Stephen Comstock
Director, Communications/Publications

Table of Contents

Acknowledgments

This design guide was produced under ASHRAE Research Project RP-849 under guidance of a Project Monitoring subcommittee chaired by Mark Fly.

This guide could not have been produced without the efforts of a large number of volunteer reviewers, who provided suggestions for the content of the guide and reviewed initial drafts. In particular, special thanks go to Chad Dorgan, Mark Fly, Mike Gallagher, Don Gatley, Ken Gillespie, Jack Harmon, and Jianshun Zhang for their detailed and thoughtful review comments.

Jack Harmon provided information on mixing boxes, coil selection, the insulation hanger, and drain trap details (Figures 4-4 and 4-9) and sample control sequences in Chapter 6. Mike Pajewski of the Trane Company provided assistance with cooling coil selection and pricing. Dave Orr provided information on duct layouts and duct insulation costs. Mark Fly provided information on DX applications.

Dan Berg, Paul Opperman, Dave Waters, and Tom Sullivan provided invaluable assistance with developing numerical examples, producing graphics, formatting equations, and producing the final page layouts.

Chapter One

Introduction

1.1 Background

Cold-air distribution systems supply air for space conditioning at reduced temperatures compared to current typical air distribution designs. These systems, sometimes referred to as low-temperature air distribution systems, typically supply air between 40°F and 50°F (4°C and 10°C). This is in contrast to most "conventional" air distribution designs, which utilize supply air at nominal temperatures between 50°F (10°C) and 59°F (15°C).

The use of cold-air distribution technology can result in lower mechanical system costs, reduced energy consumption, and improved comfort. This design guide provides designers and building owners with the basic information needed to design, install, and operate successful cold-air distribution systems. The design guide is intended to be a practical how-to manual for design practitioners. It provides comprehensive coverage of cold-air distribution design considerations, including evaluation techniques, detailed design information, construction requirements, and design examples. It can also function as an entry point into the literature for those seeking more detailed information on specific aspects of cold-air distribution design. The guide does not cover standard heating, ventilating, and air-conditioning (HVAC) design procedures in depth, but it does provide appropriate references for those readers who desire more information on these topics.

Cold-air distribution is not a new technique, even though its implementation involves some changes from today's "standard" design practice. Supply air temperatures of 40°F (4°C) and less have long been used in industrial applications for humidity control. Many residential and small commercial buildings were retrofitted for air conditioning in the 1950s with 48°F (9°C) supply air, 3.5-in. (89-mm) ducts to fit inside stud walls, and high-velocity jet diffusers to mix the cold air with room air within a short distance of entering the space. Many hospitals were designed in the 1960s with 36°F to 39°F (2°C to 4°C) primary air supplied to constant-volume room induction units.

The current "conventional" standard of 55°F (13°C) supply air evolved from the desire to maintain a space relative humidity (RH) of 50% to 60% while maximizing chilled-water supply temperature and chiller efficiency. For a typical office space with a sensible heat ratio of 0.8 to 0.9, a supply air temperature of 55°F (13°C) provides a space relative humidity of 55% to 60% at 75°F (24°C). This supply temperature also allows flexibility in the selection of cooling coils, as there are many options available for a chilled-water supply temperature of 42°F to 44°F (6°C to 7°C). However, these "standard" design parameters do not necessarily maximize comfort or minimize either first cost or operating costs. They evolved largely out of convenience and have become deeply ingrained in today's HVAC industry.

With the renewed interest in ice storage cooling systems in the 1980s, the advantages of reducing supply air temperatures became evident. With the 34°F to 39°F (1°C to 4°C) chilled fluid temperatures available from ice storage, supply air temperatures of 40°F to 49°F (4°C to 9°C) could now be easily achieved, allowing significant savings in air distribution system costs and energy consumption. Improvements in occupant comfort at reduced relative humidity levels were also recognized.

These benefits encouraged designers to reevaluate the "standard" 55°F (13°C) supply air temperature and to explore the potential of cold-air distribution. Beginning with a small number of pioneers, cold-air distribution technology was developed, refined, and brought into the mainstream. At this time, hundreds of cold-air systems are successfully operating. As facility owners become more aware of the benefits of cold-air distribution, design professionals are faced with an increasing demand to provide cold-air distribution designs. This design guide was commissioned by ASHRAE to disseminate the information that engineers need to design successful cold-air distribution systems.

The remainder of this chapter provides an introduction to the benefits and limitations of cold-air distribution, sources of further information, and a discussion of the organization of this design guide.

1.2 BENEFITS AND LIMITATIONS

The benefits of cold-air distribution include:

- reduced mechanical system costs,
- decreased floor-to-floor height requirements,
- improved comfort with lower space relative humidity,
- reduced fan energy consumption and demand, and
- increased cooling capacity for existing distribution systems.

Reductions in first cost have been an important factor driving the application of cold-air distribution. Lower supply air temperatures reduce the required supply air quantity. In turn, the size of fans and ductwork is reduced, resulting in lower mechanical system costs. In some cases, cost savings resulting from the use of cold-air distribution can offset the incremental cost of adding ice storage. Owners can then realize the operating cost savings of ice storage while paying the same or less initially as for a nonstorage system.

Smaller duct sizes can also lead to reduced floor-to-floor height requirements, with significant savings in structural, envelope, and other building systems costs. In some cases, duct sizes are reduced to the point that they can be run through trusses or beam penetrations rather than under the beams. Cold-air distribution also helps increase a designer's options in applications where space for ductwork is severely limited, such as in the renovation of historic buildings.

The lower space relative humidity maintained by cold-air distribution systems provides improved comfort and indoor air quality. Laboratory research shows that subjects feel cooler and more comfortable at lower humidities, and the air is judged to be fresher with a more acceptable air quality.

Cold-air distribution systems also offer savings in energy costs. Supply fan energy consumption can be reduced by 30% to 40% due to reductions in airflow. This reduction is primarily in expensive, on-peak energy. Where cold-air distribution is used with cool storage, any increases in cooling energy are in the form of less expensive, off-peak energy. Electric demand for supply fans is also reduced up to 40%. This demand savings adds to the reduction in cooling equipment demand when ice storage is used.

Cold-air distribution can be particularly effective in systems where loads exceed the capacity of existing fans and ductwork. By reducing supply air temperatures, owners can avoid the expense of replacing or supplementing existing air distribution equipment.

There are some misconceptions about cold-air distribution that may lead to reluctance on the part of some engineers and building owners to consider it as an option. Some examples of these unfounded misgivings include the following:

Misconception: Reduced supply air volumes will make it difficult to ensure acceptable indoor air quality.

Fact: The volume of outdoor air to be supplied to the space normally remains the same as with a higher supply temperature, although the percentage is higher. The same indoor air quality considerations must be applied to a cold-air distribution design as for a higher temperature design. As with any design, systems serving multiple spaces with varying occupant densities may require special attention to ensure adequate outside air is delivered to each space. Chapter 3 discusses these considerations in detail.

Misconception: Condensation on ducts and diffusers will be a major problem.

Fact: In most buildings, the use of cold-air distribution reduces the humidity throughout the building. At these low-dewpoint conditions, the potential for condensation is no greater than in a space of higher humidity with higher supply air temperatures. See chapter 4 for further discussion.

Misconception: Cold-air distribution can only be applied with ice storage cooling.

Fact: Cold-air distribution has been widely applied to take advantage of the 34°F (1°C) to 39°F (4°C) chilled fluid available from ice storage systems. However, the benefits of cold-air distribution can also be realized if cooling is provided from a nonstorage cooling plant. See chapter 4 for further discussion.

Misconception: Low supply air temperatures will cause uncomfortable drafts.

Fact: With proper selection of terminals and diffusers, the supply air will be almost completely mixed with room air within a short distance of the diffuser. See chapter 5 for further discussion.

There are many applications where cold-air distribution is particularly attractive. Among these are instances where

- decreased floor-to-floor height will significantly reduce the total height of a high-rise building, reducing total construction cost
- space for ducts or air-handling equipment is limited;
- reduced space humidity is desirable; or
- cooling loads have increased beyond the capacity of the existing distribution system.

Of course, there are some applications where cold-air distribution should be used with caution. These include cases where

- generation of chilled fluid at 34°F to 40°F (1°C to 4°C) is not practical,
- space relative humidity must be maintained above 40%,
- high volumes of ventilation air are required, or
- economizer cooling is available with outdoor temperatures of 45°F to 55°F (7°C to 13°C) for many hours of the year.

Such applications should be analyzed on a case-by-case basis using the principles presented in this guide to determine whether cold-air distribution is applicable. The suitability of cold-air distribution for a given project depends on an informed evaluation of all applicable technical and economic factors. The information contained in this guide provides the information and methodology needed to perform such an evaluation.

1.3 SOURCES OF INFORMATION

Additional information on cold-air distribution is available from several sources. A reference list and bibliography are provided following each chapter, and a complete list of references is included at the end of the guide. In addition, information is available from the following sources:

The American Society of Heating, Refrigerating and Air-Conditioning Engineers, Inc. (ASHRAE)
The American Society of Heating, Refrigerating and Air-Conditioning Engineers, Inc. (ASHRAE), publishes technical papers and other publications and sponsors research covering a wide range of subject matter of interest to the HVAC&R industry. Technical papers presented at society meetings are published semi-annually in the *ASHRAE Transactions*. ASHRAE's *Technical Data Bulletins* contain collections of papers pertaining

to specific subjects. A number of issues devoted to cold-air distribution have been published to date and are included in the reference list in this guide. Further information on ASHRAE publications can be obtained by contacting ASHRAE Customer Service, 1791 Tullie Circle NE, Atlanta, GA 30329; phone (404) 636-8400; fax (404) 321-5478.

The Electric Power Research Institute (EPRI)

The Electric Power Research Institute (EPRI) provides funding for research that is of interest to the electric utility industry. EPRI has funded a large portion of recent research related to cold-air distribution and has published numerous research reports on cold-air distribution. Among these publications is the *Cold Air Distribution Design Guide* (2d ed., Dorgan and Dorgan 1995), a detailed reference for designers of cold-air distribution systems. EPRI reports are available from EPRI-member utilities or by contacting the EPRI Distribution Center, 207 Coggins Dr., P.O. Box 23205, Pleasant Hill, CA 94523; phone (510) 934-4212. EPRI also supports ITSAC and the HVAC&R Center (see below).

The International Thermal Storage Advisory Council (ITSAC)

The International Thermal Storage Advisory Council (ITSAC) publishes a monthly newsletter reporting on current events in the thermal storage industry and a monthly *Technical Bulletin* with articles covering cool storage applications, case studies, research reports, and design issues, including cold-air distribution. ITSAC can be contacted at 3769 Eagle Street, San Diego, CA 92103; phone (619) 295-6267.

The EPRI HVAC&R Center

The EPRI HVAC&R Center at the University of Wisconsin-Madison manages and conducts research, provides information, and performs technology transfer activities pertaining to heating, ventilating, air-conditioning, and refrigeration technology. The Center's tasks include managing much of EPRI's research on cool storage and cold-air distribution and disseminating information on cool storage and cold-air distribution to a broad audience. The HVAC&R Center can be contacted at 150 East Gilman Street, Suite 2200, Madison, WI 53703; phone (800) 858-3774; fax (608) 262-6209.

Cool Storage and Air Distribution Equipment Vendors

Manufacturers and installers of cool storage and air distribution equipment can provide valuable information concerning the use of their products in cold-air distribution systems. While this information is typically not unbiased, discussions with several vendors can often yield useful comparisons. Manufacturers' representatives can also offer assistance in sizing and selection of components.

Electric Utility Companies
Local electric utility companies are often a good source of information on cold-air distribution. Many utilities have staff members devoted to encouraging application of cool storage and cold-air distribution in their service territories. In particular, utility contacts can provide information on applicable rate schedules and incentive programs, as well as local area cold-air distribution installations.

Trade Magazines
Trade magazines often carry articles covering noteworthy cold-air distribution applications and technical design issues related to cold-air distribution. The reference lists at the end of each chapter contain a number of citations for articles from trade magazines. Additional references may be found in the most recent issues of magazines such as *ASHRAE Journal*, *Heating/Piping/Air Conditioning*, *Consulting-Specifying Engineer*, and *Engineered Systems*.

1.4 ORGANIZATION OF THE GUIDE

The remainder of this guide is organized into the following chapters:

- **Chapter 2, Design Methodology,** provides a summary of the normal design procedure for air distribution systems and a review of fundamental air distribution system design considerations. In particular, those areas where cold-air distribution design differs from "conventional" design are highlighted and discussed.
- **Chapter 3, Ventilation, and Thermal Comfort,** discusses the factors that determine adequate comfort and indoor air quality and how they are affected by the use of cold-air distribution.
- **Chapter 4, Selection of Cold-Air Distribution Components**, provides guidance in the sizing and selection of equipment for cold-air distribution systems.
- **Chapter 5, Selection of Cold-Air Diffusion Components,** discusses the methods of selecting and sizing diffusers and terminal units.

- **Chapter 6, Control Strategies,** addresses specific control system requirements for cold-air distribution.
- **Chapter 7, Commissioning,** provides commissioning recommendations for cold-air distribution systems.
- **Chapter 8, Economic Evaluation,** provides guidance in assessing first costs and operating costs of cold-air distribution systems.
- **Chapter 9, Computer Modeling,** contains a general discussion of factors to consider when simulating the performance of cold-air systems with computer models.
- **Appendix A,** lists symbols, subscripts, and constants used in equation throughout the Guide.
- **Appendix B,** provides information on conversion between IP and SI units for those units most commonly encountered in cold air distribution design.
- **An Annotated Bibliography,** lists important references for cold air distribution design and analysis, with brief descriptions of the contents.

1.5 REFERENCES

Dorgan, C.E., and C.B. Dorgan. 1995. *Cold air distribution design guide*. EPRI TR-105604. Palo Alto, Calif.: Electric Power Research Institute.

CHAPTER TWO

DESIGN METHODOLOGY

2.1 INTRODUCTION

This section provides a summary of the typical design procedure for air distribution systems and a review of fundamental air distribution system design considerations. In particular, those areas where cold-air distribution design differs from "conventional" design are highlighted and discussed. For the purposes of this guide, cold-air distribution systems are defined as those that supply air at temperatures between 40°F and 50°F (4°C and 10°C). Most "conventional" air distribution designs for space-conditioning applications utilize supply air at nominal temperatures between 50°F and 59°F (10°C and 15°C).

As discussed in chapter 1, cold-air distribution is an attractive option for many space-conditioning applications. Cold-air distribution design departs from "standard" design practice in several areas. However, with the guidance provided in this document and the application of basic engineering principles, the design of successful cold-air distribution systems can easily be achieved.

2.2 SUPPLY TEMPERATURE SELECTION

The design air distribution supply temperature affects almost all aspects of air-conditioning system design. As the supply air temperature is reduced,

- fan size and duct size decrease,
- fan heat and fan energy decrease,
- space required for fans and ducts decreases,
- latent load and associated cooling energy may increase,

- requirements for reheating minimum supply air quantities may increase,
- cooling coil face area decreases and depth increases,
- required fluid supply temperature or DX suction temperature decreases,
- requirements for tempering or diffusing supply air increase, and
- ventilation air fraction increases, although the absolute quantity generally remains constant.

The trade-offs among these factors must be evaluated to select the appropriate temperature for a given project. The optimum supply air temperature selection will minimize first cost and operating cost, while satisfying the following conditions:

- meeting both sensible and latent cooling loads,
- providing sufficient ventilation air under all load conditions,
- providing adequate air diffusion, and
- providing acceptable acoustic performance.

The criteria for evaluating the optimum supply air temperature are likely to be different for each application. Climatic conditions, thermal and acoustic comfort requirements, functional use and occupancy patterns, and the relative importance of first cost vs. operating cost may all vary from one project to another. These factors should all be considered in selecting a system to meet an owner's objectives for a specific building.

2.3 TYPICAL HVAC DESIGN PROCESS

A detailed description of the normal design process is beyond the scope of this guide. Lorsch et al. (1993) provide a complete discussion of the HVAC design procedure. The process is subject to economic and time constraints, to availability of or preferences for certain equipment, and to modifications in design criteria as the project progresses.

The design of an HVAC system typically proceeds through the following phases (Lorsch et al. 1993; ASHRAE 1996).

Program Phase

The functional use(s) of the building, the requirements of the owner and occupants, architectural concepts, and budget limitations are typically determined during this phase. These requirements should be documented in a statement of design intent for the system, which is refined and updated as the design progresses. If cold-air distribution is to be considered for the project, it is best to advance the option at this point, so the entire design can be developed to best realize its benefits. For example, the reduction in space required for ductwork may allow a revised approach to the structural design and substantial cost savings. Reduced duct space may also be offered as a solution for architectural design concepts such as vaulted ceilings.

Schematic Design

In this stage, the designer selects and compares appropriate candidate systems to meet the functional needs of the building and develops recommendations for the owner. This phase typically involves evaluation of the benefits and trade-offs of cold-air distribution relative to other design approaches. Estimates of annual energy consumption, demand, and costs, as well as system installed costs, are developed. The level of analysis may range from simple rule-of-thumb estimates to hour-by-hour simulations with detailed cost comparisons. A realistic first-cost analysis of air-side options at this point will often show cold-air distribution to be an attractive option.

Preliminary Design or Design Development

During this phase, the mechanical designer coordinates the HVAC system design with the architectural, structural, and electrical systems to resolve potential conflicts. Heat loss, heat gain, and ventilation calculations are refined, equipment sizes and capacities are selected, and system layouts are developed. If cold-air distribution is to be introduced as an option after this work has been completed, substantial engineering effort will be required for redesign.

Final Design, Preparation of Construction Documents

In this stage, the preliminary layouts and equipment selections are finalized and the final drawings and specifications are completed. Some projects will involve preparation of alternate designs for both standard and cold-air distribution systems, with bids solicited for both options. In such cases, the primary option generally receives the most effort during the design. If cold-air distribution is offered as the alternate option, designers should ensure that all of the design considerations discussed in this guide are addressed.

Construction Phase

The design engineer's responsibilities during this phase typically include reviewing shop drawings, observing construction, and conducting or witnessing system performance tests. Contractors may require guidance to ensure that they follow the project specifications rather than their standard methods. Chapter 7 discusses construction-phase considerations for cold-air distribution systems.

Acceptance or Commissioning Phase

The term "commissioning" is often used to refer to the start-up and functional performance testing of mechanical systems. However, the ASHRAE commissioning guideline (ASHRAE 1996) outlines a much more comprehensive process for ensuring that an HVAC system is designed, installed, tested, and operated in accordance with the design intent. The acceptance phase procedures constitute an important part of this process, but there are important elements that must take place during each of the earlier phases of a project. Experience has shown that a comprehensive commissioning procedure is essential to the success of a cold-air distribution project. See chapter 7 for additional discussion of commissioning for cold-air distribution systems.

Postoccupancy Services

Designers may be called upon to assist with training operating personnel or optimizing system operation. These services are often included as part of a comprehensive commissioning process.

2.4 AIR DISTRIBUTION SYSTEM DESIGN PROCEDURE

The design of an air distribution system takes place within the context of the phases of the design process outlined above. The air distribution design procedure results in specifications for equipment such as fans, cooling coils, ductwork, insulation, terminal boxes, and diffusers. Requirements for control hardware and operating sequences must also be specified. To achieve these results, a number of specific steps must be carried out. The design procedure will not necessarily consist of the same steps in the same order for every project. Some steps may be repeated in an iterative fashion as the design proceeds through the phases outlined above. However, the following general steps typically occur in some form in every design:

- Select system configuration
- Determine space cooling loads
- Determine ventilation air requirements
- Select supply air temperature and flow rate
- Calculate cooling coil load
- Select cooling coils
- Determine cooling plant requirements

- lay out ductwork
- determine insulation requirements
- select fans
- select air distribution diffusers and terminals
- develop control sequences

Each of these steps is discussed in greater detail below.

2.4.1 SELECT SYSTEM CONFIGURATION

The overall configuration of the air-conditioning system includes the type of cooling plant, the type of air distribution system, the number and location of air-handling units, and the method of delivering supply air to the space. The selection of these configuration options may be determined by constraints imposed by the owner or architect. In many cases, a preliminary design will be developed for each of several options so the optimum configuration can be selected.

The cooling plant for a cold-air distribution system must be capable of providing coolant temperatures low enough to generate the desired supply air temperature. Many cold-air distribution systems use ice storage plants as the source of cooling, but cold-air distribution systems are also installed with nonstorage cooling plants. Distribution to air-handling-unit cooling coils is generally by chilled water or by a secondary coolant such as a glycol solution. However, cold-air distribution has also been applied with direct expansion (DX) and liquid overfeed cooling coils. The selection of cooling plants for cold-air distribution systems is discussed in more detail in chapter 4.

The most common air distribution system configuration for larger commercial space-conditioning applications is variable air volume (VAV). The majority of the discussion and design examples in this guide pertain to VAV systems, although many also apply to other system options. Configurations of air-handling units range from systems employing individual units to serve single floors or building areas to those with large units serving multiple floors or an entire building.

The reduced size of air-handling units in cold-air distribution systems can affect the selection of a configuration option. A large unit serving many floors may be more practical with cold air distribution because of its smaller footprint and the reduced size of duct chases between floors. On the other hand, floor-by-floor units using cold air distribution can be installed within a smaller mechanical room on each floor.

Supply air may be delivered directly to the conditioned space by diffusers, or it may be tempered using induction boxes or fan-powered mixing boxes. Space air diffusion considerations are discussed in detail in chapter 5.

2.4.2 DETERMINE SPACE-COOLING LOADS

A detailed discussion of the calculation of space-cooling loads is beyond the scope of this guide. This subject is covered thoroughly in chapter 26 of the *1993 ASHRAE Handbook—Fundamentals* (ASHRAE 1993), in ASHRAE's *Cooling and Heating Load Calculation Manual* (McQuiston and Spitler 1992), and in ASHRAE's *Air-Conditioning Systems Design Manual* (Lorsch et al. 1993).

Load calculation considerations that are particularly important for cold-air distribution systems include:

- accounting for heat gains from supply fans and mixing box fans and heat gains to ductwork, and
- accounting for possible increased latent loads due to infiltration and diffusion of moisture through the envelope.

These and other load calculation considerations are discussed in chapter 4.

2.4.3 DETERMINE VENTILATION AIR REQUIREMENTS

Ventilation air requirements are determined by the number of occupants and type of occupancy, as specified in *ANSI/ASHRAE 62-1989, Ventilation for Acceptable Indoor Air Quality* (ASHRAE 1989) or other applicable codes or standards. Requirements for the amount of ventilation air are normally not affected by the supply air temperature, although the proportion of ventilation air in the supply air increases as the supply temperature is lowered. The determination of ventilation air requirements for cold-air distribution systems is addressed in chapter 3.

2.4.4 SELECT SUPPLY AIR TEMPERATURE AND FLOW RATE

While the selection of the design supply air temperature is subject to a number of constraints, there will typically be a range of practical temperatures from which to choose. The designer's task is to select the supply temperature that will result in the lowest first cost and operating cost while meeting the functional requirements of the system.

The product of the room-to-supply-air temperature difference and the supply airflow rate is determined by the space-cooling load. With lower supply temperature, less supply airflow is required, reducing the first cost and operating cost of air distribution equipment. However, costs for cooling generation and heat transfer may increase. Economic analysis and the selection of supply air temperature are discussed in chapter 8.

2.4.5 CALCULATE COOLING COIL LOAD

The load on a cooling coil depends on the volume of air passing through the coil and the difference in enthalpy between the entering and leaving air conditions. Regardless of the supply air temperature, proper coil selection requires accurate calculation of the psychrometric condition of the air entering the coil. The effects of space-cooling load, system heat gains, and ventilation air must be taken into account. The use of a "standard" entering air condition for specifying cooling coil selections is particularly risky in cold-air distribution design. The calculation of cooling coil loads and the determination of cooling coil entering air conditions are discussed in chapter 4.

2.4.6 SELECT COOLING COILS

Cooling coil selection is critical to the design of a successful cold-air distribution system. Chilled-fluid temperatures, entering air conditions, and leaving air conditions are different from standard designs, so conventional experience and rules of thumb typically do not apply.

The typical recommended coil face velocity for a cold-air distribution system is 300 to 450 fpm (1.5 to 2.3 m/s). This is lower than for standard designs because more moisture is being condensed from the air, and there is a greater chance of water being entrained into the airstream and deposited downstream.

Cooling coils for cold-air distribution designs are typically eight to ten rows deep. In some cases, particularly for supply temperatures of 47 °F to 50°F (8 °C to 10°C), six-row coils can provide acceptable performance. Fin spacing of 8 to 14 fins per inch (0.3 to 0.6 fins per millimeter) is commonly used for cold-air distribution, although water carryover can be a concern at the closer fin spacings. When a high chilled-water supply-to-return temperature range is desired to minimize pumping energy, more coil surface will generally be required. A detailed discussion of coil selection issues is given in chapter 4.

2.4.7 DETERMINE COOLING PLANT REQUIREMENTS

Cooling plant parameters that affect the air distribution system design include the fluid supply temperature to the coils, the properties of the heat transfer fluid, and the design temperature range. The fluid supply temperature must be low enough to produce the required supply air temperature for a given cooling coil selection. For typical coil selections, the fluid temperature should be at least 6°F (3°C) below the desired discharge air temperature, although closer approaches can be achieved with more expensive coils.

For ice storage cooling plants, the available chilled-fluid temperature depends on the amount of ice remaining in storage and the rate at which cooling is discharged from storage. The storage system must be sized to deliver fluid at or below the design temperature for each hour of the design load profile. A detailed discussion of the sizing of ice storage cooling plants is beyond the scope of this guide. This topic is addressed in ASHRAE's *Design Guide for Cool Thermal Storage* (Dorgan and Elleson 1993).

Many ice storage systems, and nonstorage systems with fluid supply temperatures below 38°F (3.3°C), use heat transfer fluids consisting of mixtures of water with ethylene glycol or propylene glycol. Considerations for applying these fluids are discussed in chapter 4.

For nonstorage cooling plants, chillers can be selected to provide any supply temperature required for cold-air distribution. Cold-air distribution systems often operate with fluid temperature differentials of 20°F to 24°F (11°C to 13°C), compared to typical "standard" differentials of 10°F to 16°F (5°C to 9°C). Some systems have been designed with ranges as high as 30°F (17°C). This increased range can result in a reduction in pump size and pumping energy of 50% or more.

The fluid temperature range will be determined by the performance characteristics of the cooling coil(s) and the cooling plant. Designers should select coils and specify control sequences to ensure that the desired range will be achieved, as discussed in chapter 4. The cooling plant must be sized to cool the distribution fluid from the design return temperature to the required supply temperature at the appropriate flow rate.

With the reduced fluid supply temperatures used in cold-air distribution systems, additional pipe insulation may be required. Guidelines for determining insulation requirements are given in chapter 4.

2.4.8 LAY OUT DUCTWORK

Standard duct design procedures are used for cold-air distribution systems. A detailed discussion of duct-sizing methods is beyond the scope of this guide. Additional information is given in chapter 32 of the *1993 ASHRAE Handbook—Fundamentals* (ASHRAE 1993).

The designer has a number of options in sizing ductwork for cold-air distribution systems. Just as for standard designs, tradeoffs must be made between duct size on one hand and total pressure requirements and fan energy on the other. If the same pressure loss criteria are used as for traditional designs, duct size and fan energy will be reduced in proportion to the reduction in air volume. Duct size can be reduced even further if higher pressure losses are acceptable. At the other extreme, air velocities and fan energy can be reduced to very low levels if "standard" duct sizes are used with reduced supply air temperatures.

Where the available ceiling space limits the height of ducts, the use of cold-air distribution allows designers to select more favorable duct aspect ratios. In some cases, it will be practical to use round or oval ductwork, which provides lower pressure losses, improved acoustic performance, and lower first cost. The reduction in fan and duct size may also allow designers to avoid undesirable fan entry and exit sections, which result in system effects that impair fan performance.

2.4.9 DETERMINE INSULATION REQUIREMENTS

Insulation on ductwork and other air distribution equipment serves to reduce heat gain to the supply air and to prevent condensation of moisture from the ambient air. Insulation takes on increased importance in cold-air distribution systems. The lower supply air temperature results in an increased temperature difference relative to the surroundings, increasing the heat gain. The lower supply temperature also increases the possibility of some surfaces being cooled below the ambient dew point and causing condensation.

Duct insulation thickness in cold-air distribution systems is commonly increased to minimize heat gains. Where 1-inch (25-mm) insulation would be used with a 55°F (13°C) design, a 2-inch (51-mm) thickness is often specified for cold-air systems. Thorough coverage; a complete, unbroken vapor barrier; and good workmanship are also especially important. Chapter 4 provides additional information and a detailed discussion of insulation requirements.

2.4.10 SELECT FANS

Fan selection for cold-air distribution is similar to that for "standard" systems. One significant consideration is the placement of the cooling coil relative to the fan. In a blow-through configuration, the fan is upstream of the cooling coil, while a draw-through configuration places the fan downstream of the coil. These approaches differ in how the fan heat is removed by the cooling coil. Chapter 4 provides a discussion of fan placement issues.

2.4.11 SELECT AIR DISTRIBUTION DIFFUSERS AND TERMINALS

The delivery of 40°F to 50°F (4°C to 10°C) supply air directly to the space raises concerns about the ability of diffusers to adequately mix the supply air with room air before it reaches the occupied zone. The possibility of condensation on diffuser surfaces is also a concern. To avoid these issues, many cold-air distribution systems have been designed with fan-powered mixing boxes. Fan-powered boxes mix recirculated room air with cold primary air before delivering it to the diffusers. With this approach, the air distribution system downstream of the boxes is equivalent to a standard 55°F (13°C) design. However, because of the small, relatively inefficient fans and motors typically used in fan-powered boxes, the energy consumption of the boxes offsets any fan energy savings arising from the reduction in supply air temperature (Elleson 1993).

Recent research has clarified the issues related to supplying cold air directly from diffusers and has led to the development of guidelines for selecting diffusers for cold-air applications. Chapter 5 provides a detailed discussion of selecting diffusers and terminal boxes for cold-air distribution systems.

2.4.12 DEVELOP CONTROL SEQUENCES

Designers should begin defining control sequences early in the design development as an integral part of defining the system design intent. Cold-air distribution system designs may include specific control sequences for mixing boxes, for resetting supply air temperatures, and for humidity control when starting up after an overnight or weekend shutdown period. Standard economizer control sequences may also need to be modified based on the temperatures usable for free cooling. Chapter 6 discusses these and other control sequences.

2.5 CONSIDERATIONS FOR COLD-AIR DISTRIBUTION RETROFIT

Cold-air distribution is often recommended if space-cooling loads on an existing system exceed the distribution system capacity. The design procedure in such cases is similar to a design for new construction. However, the designer is generally limited by the performance characteristics of the existing equipment and has less flexibility in the selection of design parameters.

The design process for retrofit applications generally includes the following steps, in addition to the considerations for new construction.

- Calculate the current or expected design loads. If available, measurements of design loads, on a zone level or a system level, are preferable to calculations. If loads cannot be measured at design conditions, designers can use near-design or part-load measurements to supplement load calculations. Note that if loads exceed the capacity of the existing equipment, load measurements should be used with care. In such cases, measured loads will be limited by equipment capacity and may not reflect the actual loads.
- Determine the required supply air temperature to meet the design loads with existing fans and ductwork.
- Evaluate whether existing duct insulation will prevent condensation on ducts and limit duct heat gain to acceptable levels. Determine what, if any, additional insulation is require and whether special measures will be needed to avoid thermal bridging through insulation.
- Determine whether existing diffusers can provide acceptable performance at the new supply air temperature. Evaluate performance at part-load conditions. Develop temperature reset operating sequences if necessary. If expected performance is not acceptable, consider replacing diffusers or adding induction or fan-powered boxes.
- Evaluate the ability of the existing cooling coils to cool the required air volume to the required temperature. Determine whether the existing coils will need to be replaced or supplemented by additional coils. Verify that space is available for new coils.

- Determine the required chilled-fluid temperature and flow to each of the coils. Determine the diversified block load and the total chilled-fluid flow requirement.
- Evaluate the ability of existing chillers to deliver the required chilled-fluid flow at the required temperature. Determine the required additional chiller or ice storage capacity.

2.6 PROTOTYPE SYSTEM FOR DESIGN EXAMPLES

To illustrate the design considerations and calculations required for cold-air distribution design, the following chapters provide numerical examples illustrating many cold-air distribution design considerations and calculations.

The examples illustrate the steps of the cold-air distribution design procedure for a system serving one floor of a multistory commercial office building. The space includes 15 zones of general office occupancy and one conference room with a higher occupant density. The total conditioned area of the space is 16,650 ft^2 (1,550 m^2). Figure 2-1 shows the zone layout and dimensions of the space.

The total occupancy of the space is assumed to be 117 people, based on the recommendation of 7 people per 1,000 ft^2 (93 m^2) given in *ANSI/ASHRAE Standard 62-1989* (ASHRAE 1989).

The space temperature setpoint for both cold-air distribution and conventional examples is 75°F (24°C). The space relative humidity is initially taken as 50% for the conventional design and 40% for the cold-air distribution design. These humidity values are verified by the psychrometric analysis presented in section 4.10, based on the results of other calculation steps.

The examples use design outdoor conditions of 91°F db, 74°F wb (33°C db, 23 °C wb), typical of the ASHRAE 1% design conditions for many locations in the United States.

The calculated block sensible load for the space is 350,000 Btu/h (102,600 W). Latent loads are calculated in section 4.3. Other data pertinent to the specific examples are presented within the respective sections.

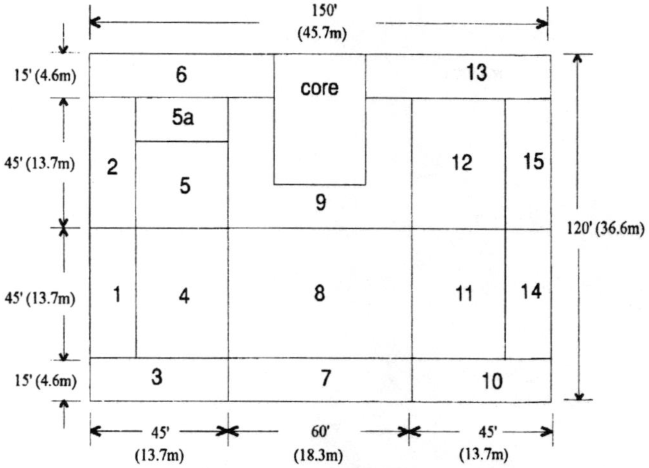

FIGURE 2-1 Zone layout for example building.

2.7 REFERENCES

ASHRAE. 1989. *ANSI/ASHRAE Standard 62-1989, Ventilation for acceptable indoor air quality.* Atlanta: American Society of Heating, Refrigerating and Air-Conditioning Engineers, Inc.

ASHRAE. 1993. *1993 ASHRAE handbook—Fundamentals.* Atlanta: American Society of Heating, Refrigerating and Air-Conditioning Engineers, Inc.

ASHRAE. 1996. *ASHRAE Guideline 1-1996, The HVAC Commissioning Process.* Atlanta: American Society of Heating, Refrigerating and Air-Conditioning Engineers, Inc.

Dorgan, C.E., and J.S. Elleson. 1993. *Design guide for cool thermal storage.* Atlanta: American Society of Heating, Refrigerating and Air-Conditioning Engineers, Inc.

Elleson, J.S. 1993. Energy use of fan-powered mixing boxes with cold air distribution. *ASHRAE Transactions* 99(1): 1349-1358.

Lorsch, H.G., et al. 1993. *Air-conditioning systems design manual.* Atlanta: American Society of Heating, Refrigerating and Air-Conditioning Engineers, Inc.

McQuiston, F.C., and J.D. Spitler. 1992. *Cooling and heating load calculation manual,* 2d ed. Atlanta: American Society of Heating, Refrigerating and Air-Conditioning Engineers, Inc.

Chapter Three

Ventilation and Thermal Comfort

3.1 Introduction

The HVAC systems of commercial and residential buildings are designed to ensure both adequate ventilation and occupant comfort. The volume flow rate of air through the HVAC system of a building is found from the largest of either the ventilation requirement or the sensible heating/cooling load. The ventilation requirements usually govern the minimum supply air conditions in variable-air-volume systems at low sensible and latent load. The difference between the ventilation flow rate and the cooling/heating flow rate is met by reconditioning of the return air.

The consideration of ventilation relates to the health of the occupants, which is protected by adequate ventilation of the building with outdoor air. This is needed since indoor air can be contaminated by emissions from occupants, processes, building materials, and furnishings. The emissions result in elevated levels of contaminants, such as carbon dioxide, volatile organic compounds (VOCs), particulate matter, inorganic gaseous combustion products such as CO and NO, radon progeny, formaldehyde, pesticides, and aerosols. The indoor contamination is removed by either a mixing or displacement ventilation approach. In this Guide, the discussion is centered on the mixing ventilation approach. With mixing ventilation, contaminated indoor air is diluted with cleaner outdoor air brought in by the HVAC system. The mixing and dilution take place at the central air handler and near the supply jet outlet in the unoccupied zone of the room. A portion of the diluted gases is exhausted from the building, and the unexhausted gases are returned to the main air handler.

The occupants' thermal comfort is achieved by controlling the thermal environment within given ranges of temperature, humidity, and air speed. The recommended ranges of these variables for a variety of activity levels are given by *ANSI/ASHRAE Standard 55-1992* (ASHRAE 1992) and are discussed in section 3.5 of this guide and in chapter 8, "Physiological Principles, Comfort, and Health," of the *1993 ASHRAE Handbook—Fundamentals* (ASHRAE 1993).

Constant-volume ventilation systems vary the supply air temperature and keep the flow rate of supply air constant. In a constant-volume system without an economizer cycle, the amount of outdoor air brought into the building is constant at all times. The outdoor and return dampers are set at the desired minimum position, so the supply volume and the percentage of outdoor air do not change. Outdoor air can be delivered separately to each occupied zone using a number of independent constant-volume systems.

Variable-air-volume (VAV) systems meet changes in the space cooling/heating load by varying the supply airflow rate. In general, the VAV system maintains the supply air at a constant supply temperature and reduces or increases the airflow between preset maximum and minimum flows to meet the space loads. The supply air temperature at the cooling coil is typically set at 55°F (13°C) for conventional systems and at 45 °F (7.2°C) for cold-air systems. It is also possible to reset the supply air temperature in a VAV system to prevent overcooling of a space. This is discussed in section 6.4 of this guide.

The flow rate through each VAV terminal is controlled by a zone thermostat. The terminal is fully open at the maximum zone load. As the load decreases, the zone control thermostat will cause the VAV damper to begin to close and increase the pressure drop across the damper. This increases the duct pressure, which produces a reduction in system airflow. The throttling ratio, the ratio of the actual flow rate to the design flow rate, depends on the ratio of the ventilation and cooling requirements. The system throttling ratio will not be equal to the zone throttling ratios due to the different loads and ventilation requirements. Fan bypass, discharge dampers, inlet vanes, or variable-speed fan control can be used for duct pressure control.

Building air-conditioning systems using VAV need to be designed to maintain the ventilation rate under varying load conditions. Filardo (1993) discusses the VAV performance for multiple spaces under variable load conditions and the minimum outdoor air required to maintain acceptable indoor air quality using the Standard 62 (ASHRAE 1989) multiple-space equation.

The exhaust air fan will cause outdoor air to infiltrate unless outdoor air fans supply a quantity of outdoor air equal to the exhausted plus exfiltrated air. Buildings are usually pressurized to reduce or prevent infiltration. It is customary to supply approximately 5% to 20% more outdoor air than exhaust air to ensure exfiltration.

3.2 VENTILATION ACCORDING TO ASHRAE STANDARD 62-1989

ANSI/ASHRAE Standard 62-1989, Ventilation for Acceptable Indoor Air Quality, defines the requirements for minimum ventilation rates (ASHRAE 1989). The Ventilation Rate Procedure given in the standard prescribes the rate at which ventilation air must be delivered to a given space. The discussion in this guide is based on the use of the Ventilation Rate Procedure. This method is the most common method used to determine ventilation rates for a system under design.

An alternative method, the Indoor Air Quality Procedure, provides guidelines for the specification of acceptable concentration of certain contaminants in indoor air, but it does not prescribe ventilation rates or air treatment methods. This method may be used in situations where the Ventilation Rate Procedure does not provide appropriate results. For example, the Indoor Air Quality Procedure may be used

- if the ventilation rate determined by the Ventilation Rate Procedure is not adequate to control a specific contaminant or
- if all contaminant concentrations can be maintained at acceptable levels with lower ventilation rates than called for by the Ventilation Rate Procedure.

The ventilation requirements for commercial and institutional buildings are given in Standard 62-1989. These ventilation rates were chosen to achieve an acceptable level of indoor air quality by control of carbon dioxide, particulates, odors, and other contaminants common to those spaces. If cleaned, recirculated air is used to reduce the outdoor air below the recommended values, the Indoor Air Quality Procedure must be used. The requirements are given in terms of quantity of ventilation air per person or per unit of floor area. If the actual number of occupants in a space is not known, the table provides typical occupancies in terms of people per 1,000 ft^2 or per 100 m^2.

Office spaces and conference rooms require 20 cfm (10 L/s) of outdoor air per person. Office spaces have an estimated maximum occupancy density of 7 people per 1,000 ft^2 or per 100 m^2, and conference rooms have an estimated maximum occupancy density of 50 people per 1,000 ft^2 or per 100 m^2, respectively.

3.3 OUTDOOR AIR FRACTION WITH COLD AIR

The supply airflow rate for a known space load and diffuser outlet temperature is found using the energy equation:

$$Q_s = \frac{q}{\{C_1*(T_{room} - T_o)\}} \tag{3-1}$$

where

Q_S = supply airflow rate, cfm (L/s)

q = space load, Btu/h (W)

C_1 = 1.1 Btu·min/ft^3·h·°F (1.23 W·s/m^3·°C)

T_{room} = average space temperature, °F (°C)

T_O = diffuser outlet temperature, °F (°C)

The required outdoor airflow rate is determined by the occupancy and the Standard 62 ventilation requirement:

$$Q_{oa} = \sum_i N_i * R_i \tag{3-2}$$

where

Q_{oa} = sum of the outdoor air required by all zones, cfm (L/s)

N_i = number of occupants in zone i

R_i = required ventilation rate per occupant in zone i, cfm (L/s)

The outdoor air fraction, X, is the ratio of the outdoor airflow rate to the supply airflow rate:

$$X = \frac{Q_{oa}}{Q_s} \tag{3-3}$$

The outdoor air fraction with cold-air systems will increase relative to that of conventional systems since the supply airflow rate is decreased as the supply air temperature is decreased, and the outdoor airflow must remain constant to meet ventilation requirements. If the minimum outdoor airflow rate required is greater than the supply airflow rate, the supply air temperature will need to be reset to a higher temperature to meet the minimum outdoor air requirements. This situation can occur with a low cooling load accompanied by a high occupant density.

3.4 VENTILATION OF MULTIPLE SPACES

3.4.1 MULTIPLE-SPACE EQUATIONS

With multiple spaces served by a common supply system, the required outdoor air fraction in each space may differ from zone to zone, since the ventilation requirement for each zone may differ. A given zone may have a required outdoor air fraction greater or less than the outdoor air fraction of the central system.

Standard 62-1989 has a method by which to reduce the required outdoor air fraction by using "unused" outdoor air from zones with lower occupant density. The standard specifies a multiple-space procedure using a two-space model. One space is the critical zone with the highest ventilation load. The other space represents the sum of the other zones that require a smaller fraction of outdoor air than the critical zone.

To deliver the required outdoor air fraction to the critical zone, the outdoor air fractions in the supply duct and in the return duct need to be considered. If the supply contains the outdoor air fraction needed to ventilate the critical zone, the other zones will be overventilated, so a fraction of the return air can be considered as a supply air source. This will reduce the outdoor air fraction in the supply air, denoted by the symbol *Y*, to a value less than the outdoor air required for the critical zone, denoted by *F*. The Standard 62-1989 equations that take the above considerations into account are:

$$F = \frac{Q_{coa}}{Q_{cs}} \tag{3-4}$$

$$Y = \frac{X}{(1 + X - F)} \tag{3-5}$$

$$Q_{oa-req} = Y^* Q_s \tag{3-6}$$

where

F = outdoor air fraction required by critical zone;

Q_{coa} = outdoor air required by the critical zone, cfm (L/s);

Q_{cs} = supply air to the critical zone, cfm (L/s);

$Q_{oa\text{-}req}$= total outdoor air required, cfm (L/s);

X = outdoor air fraction; and

Y = corrected outdoor air fraction.

The impact of Equation 3-5 on the outdoor airflow rates will be minimized if the supply air to the critical space is increased using fan-powered boxes transferring air from a common return air plenum (Ke and Mumma 1996). An alternative solution is to use a separate air distribution system to meet the critical space ventilation requirements.

3.4.2 EXAMPLE: VENTILATION OF MULTIPLE SPACES

The dependence of a critical zone outdoor air fraction (F), supply air volume (Q_{cs}), and corrected ventilation air volume ($Q_{oa\text{-}req}$) on the outlet temperature is computed in the example in this section. In this example, the conference room is the critical zone for the commercial office building example of section 2.6. The overall load, q, for the building floor is 350,000 Btu/h (102,575 W). The total required outdoor airflow for the 117 occupants, Q_{oa}, is 3,540 cfm (1,670 L/s). The conference room load, q_c, is 15,354 Btu (4,500 W). The conference room required outdoor airflow, Q_{coa}, is 450 cfm (212 L/s).

For the case of 45°F (7.2°C) air leaving the coil, the supply air temperature at the diffuser, T_o, is typically about 50°F (8.6°C). The calculations shown below indicate that by including the unused ventilation air, an overall ventilation fraction of 0.60 providing 7,636 cfm (3,270 L/s) of outdoor air will meet the required 0.81 fraction in the critical zone. The design data are as follows:

- q = 350,000 Btu/h (102,579 W)
- q_c = 15,354 Btu/h (4,500 W)
- T_{room} = 75°F (23.9°C)
- T_o = 50°F (8.6°C)
- Q_{oa} = 3,540 cfm (1,670 L/s)

From Equation 3-1, the supply airflow rate is:

$$Q_s = \frac{q}{\{1.1*(T_{room} - T_o)\}}$$

$$Q_s = \frac{350,000}{\{1.1*(75-50)\}} = 12,727\ cfm \qquad \left(Q_s = \frac{102,579}{\{1.23*(23.9-8.6)\}} = (5,450\text{L/s})\right)$$

The critical zone supply flow rate is:

$$Q_{cs} = \frac{q}{\{1.1*(T_{room} - T_o)\}}$$

$$Q_{cs} = \frac{15,354}{\{1.1*(75-50)\}} = 558\text{ cfm} \qquad \left(Q_{cs} = \frac{4,500}{\{1.23*(23.9-8.6)\}} = (239\text{L/s})\right)$$

From Equation 3-3, the building outdoor air fraction is

$$X = \frac{Q_{oa}}{Q_s} = \frac{3,540}{12,727} = 0.29 \qquad \left(= \frac{1670\text{ L/s}}{5450\text{ L/s}} = 0.29\right)$$

From Equation 3-4, the critical zone supply air fraction is

$$F = \frac{Q_{coa}}{Q_{cs}} = \frac{450}{558} = 0.81 \qquad \left(= \frac{212\text{ L/s}}{263\text{ L/s}} = 0.81\right)$$

From Equation 3-5, the corrected outdoor air fraction is

$$Y = \frac{X}{(1+X-F)} = \frac{0.29}{(1+0.29-0.81)} = 0.60$$

From Equation 3-6, the corrected outdoor airflow rate is

$$Q_{oa-req} = YQ_s = (0.60)(12,727) = 7,636\text{ cfm} \qquad \left(= (0.60)(5,450) = (3,270\text{ L/s})\right)$$

The above calculations were repeated for diffuser outlet temperatures ranging from 40°F (4.4°C) to 60°F (15.6°C) and plotted in Figures 3.1 and 3.2. The change in the supply air and corrected outdoor air volume is plotted in Figure 3.1 as a function of the outlet temperature. As the supply air temperature is lowered from 60°F to 44°F (16°C to 6.7°C), the total supply airflow rate decreases and, at outlet temperatures below 44°F (6.7°C), the supply airflow rate is less than the required ventilation air. Since the actual flow rate should always meet or exceed the required ventilation air, the supply air temperature for this load condition should not be lower than 44°F (6.7°C).

The dependence of the ventilation fractions on the diffuser outlet temperatures is plotted in Figure 3-2. As the outlet temperature is

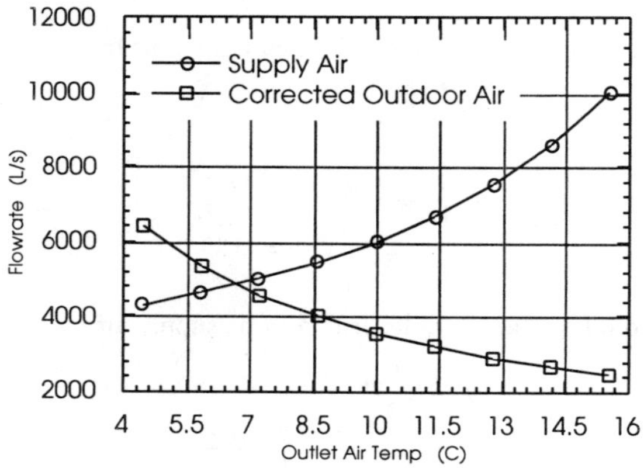

FIGURE 3-1 (SI) Airflow versus outlet air temperature.

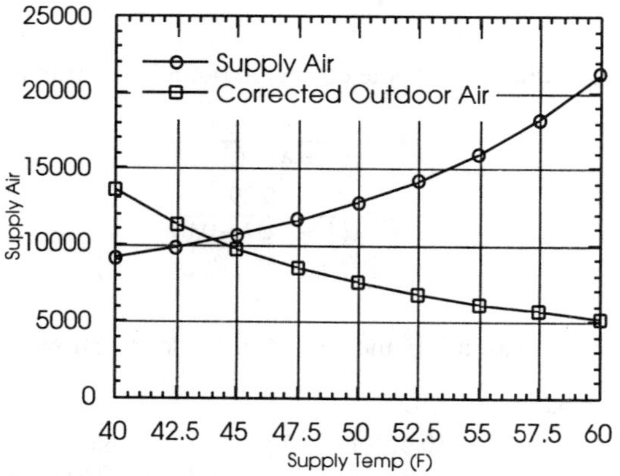

FIGURE 3-1 (IP) Airflow versus outlet air temperature.

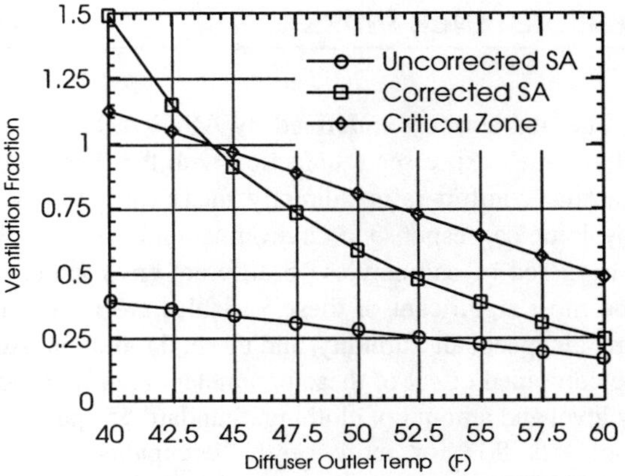

FIGURE 3-2 (IP) Ventilation fractions versus outlet temperature.

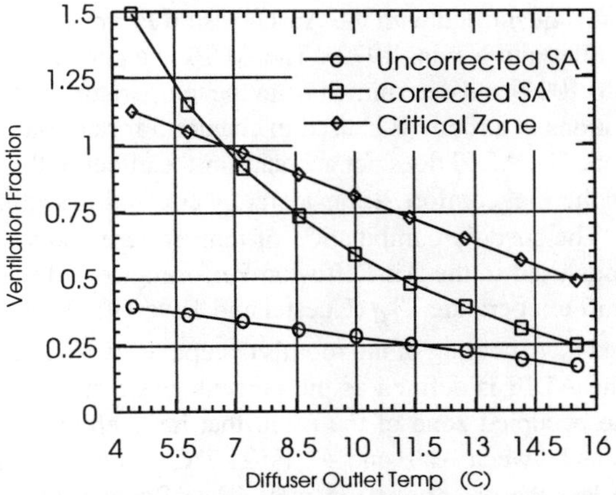

FIGURE 3-2 (SI) Ventilation fractions versus outlet temperature.

lowered from 60°F to 44°F (16°C to 6.7°C), the uncorrected supply air fraction increases from 0.20 to 0.40, the critical zone supply air fraction increases from 0.25 to 1.00, and the corrected supply air fraction increases from 0.25 to 0.99. This example shows that for cases with differing occupant densities in zones served by a single system, the required outdoor air volume increases as the supply air temperature is reduced. This effect can be mitigated by using fan powered mixing boxes to transfer air from a common return to the critical space (ASHRAE 1995; Ke and Mumma, 1996).

3.5 THERMAL COMFORT AND THE AIR DIFFUSION PERFORMANCE INDEX

Thermal comfort is defined by ASHRAE as "that condition of mind which expresses satisfaction with the thermal environment." Thermal comfort is not directly measurable but is an individual physiological response to environmental conditions. This response is triggered by as many as 15 different environmental parameters. The most significant of these variables are air dry-bulb temperature, air speed, air humidity, and thermal radiation. Acceptability of the combined effect of these parameters is influenced by the activity level and amount of clothing. Standard 55 specifies those conditions that 80% or more of the occupants will find thermally acceptable. A comprehensive overview and history of room thermal comfort testing and criteria is given in Fountain (1991).

The comfort performance of diffusers, including cold-air diffusers, is quantified with the Air Diffusion Performance Index (ADPI) (Miller and Nevins 1972). The ADPI is a percentage varying from 0 to 100% and determined from the temperature and velocity distributions in a room. It is used in chapter 5 for cold-air diffuser selection. The ADPI does not account for the effect of the humidity level on thermal comfort, a consideration covered in section 3.6

The specific combination of temperature and velocity distributions used in the Air Diffusion Performance Index is the effective draft temperature, T_{ed} (Koestel and Tuve 1955). Draft is the undesired local cooling of the room's occupants caused by air movement. The ADPI is defined as the percentage of measurement points in the occupied zone of the room that have effective draft temperatures between −3°F and +2°F (−1.7°C and +1.1°C) and air speeds of less than 70 fpm (0.35 m/s). The effective draft temperature is the difference between the local air temperature and the temperature of room air that would give the same cooling effect as local air movement. The equation for the effective draft temperature (Rydberg and Norback 1949) is

$$T_{ed} = (T - T_{avg}) - a(V - b) \tag{3-7}$$

where

T_{ed} = effective draft temperature, °F (°C)

T = local temperature, °F (°C)

a = control constant, $a = 0.07$ (8.0)

b = control velocity, $b = 30$ fpm (0.15 m/s)

T_{avg} = occupied zone average temperature, °F (°C)

V = local air speed, fpm (m/s)

An effective draft temperature below −3 °F (−1.7 °C) would be perceived as cold, one above +2°F (1.1°C) would be perceived as too hot, and local velocities above 70 fpm would be perceived as too drafty. For example, if the local temperature was 70°F (21°C) with an occupied zone average temperature of 72°F (22.2°C) and a local air speed of 50 fpm (0.25 m/s), the effective draft temperature would be −3.4°F (−1.9°C), a condition that would be perceived as too cold.

3.6 EFFECT OF LOWER RELATIVE HUMIDITY ON PERCEIVED AIR QUALITY

HVAC systems with cold-air distribution will operate normally at a lower relative humidity and dew-point temperature than conventional systems. As the supply air temperature is lowered, more water will condense out of the airstream at the cooling coil, lowering the humidity ratio. When supply air with a low humidity ratio is mixed with room air, the space will have a lower moisture content, resulting in lower relative humidity and dew-point temperatures.

As shown in Table 3-1, representative relative humidities of cold-air systems are about 40%, which is about 10% lower than the representative 50% of a conventional system. Determination of the space relative humidity and dew point for a specific example are given in section 4.10.

TABLE 3-1 REPRESENTATIVE COLD-AIR AND CONVENTIONAL ROOM CONDITIONS

Space Conditions	Room Conditions		
	Dry-Bulb Temperature °F (°C)	Relative Humidity (%)	Dew Point °F (°C)
Conventional	75 (23.9)	50	55.1 (12.8)
Cold Air	75 (23.9)	40	49.1 (9.5)

The effect of the relationship between the relative humidity and the dew-point temperature is plotted in Figure 3.3 for a constant space dry-bulb temperature of 75°F (23.9°C). As the relative

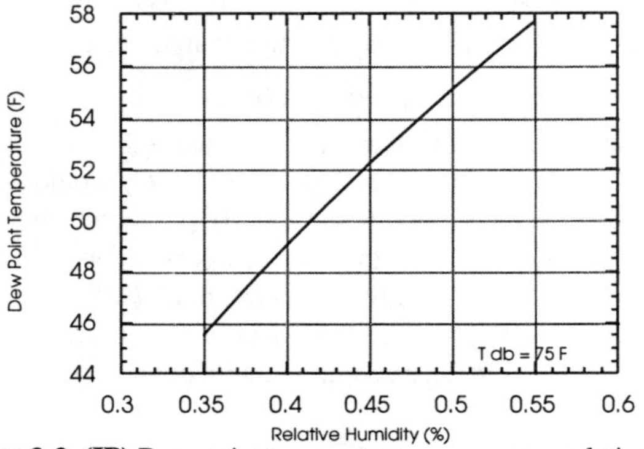

FIGURE 3-3 (IP) Dew-point temperature versus space relative humidity.

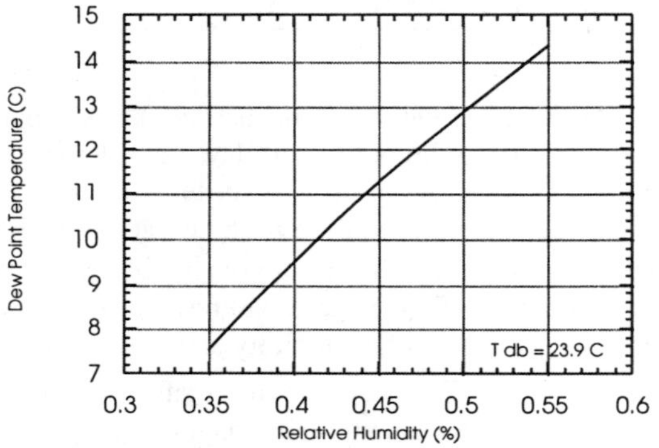

FIGURE 3-3 (SI) Dew-point temperature versus space relative humidity.

humidity is decreased from 55% to 35%, the dew-point temperature decreases from 57°F to 46°F (14.1°C to 7.5°C).

The lower level of relative humidity has a significant effect on the thermal comfort in the space. Studies by Berglund and Cain (1989), Berglund (1991), and Berglund (1994) have shown that air is perceived to be fresher and less stuffy as the space humidity is lowered. For a fixed temperature, as the humidity is decreased, air is perceived to be fresher and less stale.

Berglund (1991) reports that a 10.5°F (5.8°C) change in dew-point temperature will have the same effect on thermal sensation as a 1°F (0.5°C) dry-bulb temperature change. The increased comfort at lower humidity levels is due to reduced skin moisture and perspiration. In a warm environment but one with a low humidity level, a person will experience little discomfort if the perspiration evaporates immediately and the skin remains dry. The friction between the skin and clothing also decreases at lower humidity levels, so that fabrics feel smoother and clothing is less sticky.

Cold-air systems should be designed for the same space temperature setpoint as conventional systems so that they will provide a greater thermal comfort level than conventional systems. Additional energy savings may be achieved in operation if the space setpoint can be increased. These results indicate that it is possible to increase the dry-bulb temperature for low-humidity cold-air systems, if desired, to reduce cooling loads and still maintain the same thermal comfort level.

Using the equivalent of a 1°F (0.5°C) dry-bulb temperature increase to a 10.5°F (5.8°C) decrease in dew-point temperature, the corresponding relation between an increase in dry-bulb temperature and a decrease in relative humidity is plotted in Figure

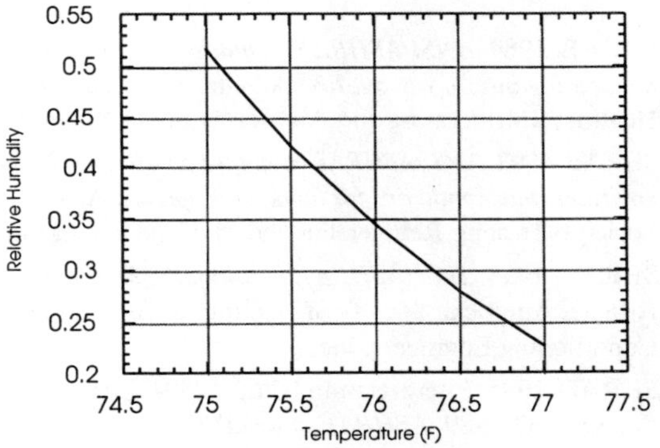

FIGURE 3-4 (IP) Equivalent thermal comfort.

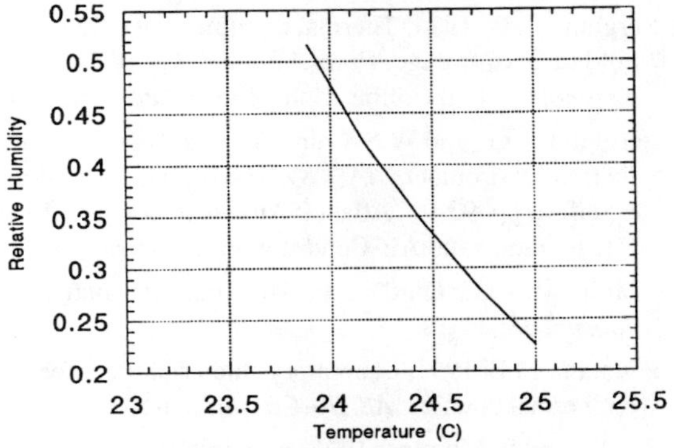

FIGURE 3-4 (SI) Equivalent thermal comfort.

3-4. Figure 3-4 shows that it is possible with a cold-air distribution system to increase the dry-bulb temperature from 75°F to 76°F (23.9°C to 24.4°C) if the relative humidity is decreased from about 50% to 35%, while maintaining an equivalent comfort condition.

There is a lower limit to space relative humidity levels. Very low humidity, below about 30%, can negatively affect comfort and health. Low humidity can lead to drying of the skin and mucous surfaces. Comfort complaints about dry nose, throat, eyes, and skin occur in situations where the dew point is less than 40°F (4°C).

3.7 REFERENCES

ASHRAE. 1989. *ANSI/ASHRAE Standard 62-1989, Ventilation for acceptable indoor air quality*. Atlanta: American Society of Heating, Refrigerating and Air-Conditioning Engineers, Inc.

ASHRAE. 1992. *ANSI/ASHRAE Standard 55-1992, Thermal environmental conditions for human occupancy*. Atlanta: American Society of Heating, Refrigerating and Air-Conditioning Engineers, Inc.

ASHRAE. 1993. *1993 ASHRAE handbook—Fundamentals*. Atlanta: American Society of Heating, Refrigerating and Air-Conditioning Engineers, Inc.

ASHRAE. 1995. Interpretation IC 62-1989-21 of ANSI/ASHRAE Standard 62-1989. *ASHRAE Journal* 37(9): 62-63.

Berglund, L.G. 1991. Comfort benefits for summer air conditioning with ice storage. *ASHRAE Transactions* 97(1): 843-847

Berglund, L.G. 1994. Thermal comfort with cold air distribution/cold thermal storage. *Proc. Advances in Cold Air Distribution Technology*. Fort Collins, Colo.: Electric Power Research Institute.

Berglund, L.G., and W.S. Cain. 1989. Perceived air quality and the thermal environment. *IAQ '89*, The Human Condition: Health and Comfort. pp. 93-99, Atlanta: American Society of Heating, Refrigerating and Air-Conditioning Engineers, Inc.

Filardo, M. 1993. Outdoor air: How much is enough? *ASHRAE Journal* 35: 34-38.

Fountain, M.E. 1991. Laboratory studies of the effect of air motion on thermal comfort. *ASHRAE Transactions* 97(1): 863-873.

Ke, Y., and S. Mumma. 1996. A generalized multiple-spaces equation to accommodate any mix of close off and fan-powered VAV boxes. *ASHRAE Transactions* 102(1).

Koestel, A., and G.L. Tuve. 1955. Performance and evaluation of room air distribution systems. *ASHRAE Transactions* 61: 533.

Miller, P.L., and R.G. Nevins. 1972. An analysis of the performance of room air distribution systems. *ASHRAE Transactions* 78(1): 191.

Rydberg, J., and P. Norback. 1949. Air distribution and draft. *ASHVE Transactions* 55: 225.

CHAPTER FOUR

SELECTION OF COLD-AIR DISTRIBUTION COMPONENTS

This chapter presents information and guidelines for sizing and selecting the components used in cold-air distribution systems. This chapter considers the cooling plant, fan, cooling coil, and ductwork. Terminal units, VAV boxes, and diffusers are covered in chapter 5.

The selection of components for a cold-air distribution system is governed primarily by the desired supply air temperature. In turn, the optimum supply air temperature depends on the cost, performance, and energy efficiency of the available equipment.

4.1 STANDARD AIR CALCULATIONS

The discussion, calculations, and examples presented in this Guide are generally in terms of volume flow rates based on standard air properties. The density of moist air varies with local barometric pressure (which varies with altitude), temperature, and moisture content. Accurate calculation of energy transfer and psychrometric properties requires that flow rates be specified in terms of mass rather than volume. However, common design practice uses engineering formulas based on volume flow rates, and volume values are often required for selection of coils, fans, ducts, and other components.

This issue is typically resolved by expressing airflow volumes in terms of standard airflow. Standard air is defined as dry air at 70°F and 14.696 psia, with a density of 0.075 lb/ft^3 (20°C and 101.325 kPa absolute, with a density of 1.204 kg/m^3). When the air density is significantly different from standard, the actual flow rate is adjusted to the mass-equivalent standard flow rate, according to the ratio of the actual density to the standard density.

$$Q_{std} = Q_{act}\frac{\rho_{act}}{\rho_{std}} \quad (4\text{-}1)$$

or

$$Q_{std} = Q_{act}\frac{v_{std}}{v_{act}} \quad (4\text{-}2)$$

where

Q_{std} = standard airflow rate, scfm (L/s)

Q_{act} = actual airflow rate, acfm (L/s)

ρ_{std} = standard air density, 0.075 lb/ft^3 (1.2 kg/m^3)

ρ_{act} = actual air density, lb/ft^3 (kg/m^3)

v_{std} = standard specific volume, 13.33 ft^3/lb (0.833 m^3/kg)

v_{act} = actual specific volume, ft^3/lb (m^3/kg)

The air density is calculated from standard density using the ratio of the absolute temperature at standard conditions, 70°F (21°C) or 530°R (294 K), to the absolute temperature of the supply air.

$$\rho_{act} = \rho_{std}\frac{T_{std}}{T_{act}} \quad (4\text{-}3)$$

where

ρ_{std} = 0.075 lb/ft^3 (1.2 kg/m^3)

T_{std} = 530°R (294 K)

T_{act} = T_{act} (°F) + 460 (Tact [°C] + 273)

For most situations typically encountered in cold-air distribution system design, the use of actual flow rates with the equations presented in this Guide will provide acceptable accuracy. Exceptions include applications at altitudes higher than 2,000 ft (600 m) above sea level or temperatures above 100°F (38°C). Engineers designing projects at high altitudes should convert actual flow rates to standard values or make the appropriate adjustments to the factors presented here to account for local barometric pressure. Designers should also size outdoor air intake ducts, dampers, and louvers for the maximum actual volume flow rate, which will be higher than the standard flow rate at summer design conditions.

Psychrometric calculations on a mass flow basis are discussed in chapter 6 of the *1993 ASHRAE Handbook—Fundamentals* and in

ASHRAE's *Cooling and Heating Load Calculation Manual* (McQuiston and Spitler 1992, chapter 10 and Appendix D).

4.2 COOLING PLANT

Cold-air distribution systems commonly use ice storage systems or nonstorage plants as the source of cooling. These systems distribute chilled water or a secondary coolant to remote air-handling units. However, cold-air distribution is also applied with packaged or built-up direct refrigeration air-handling equipment, in which the cooling coil is the evaporator for the refrigeration system.

A chiller or ice storage cooling plant must be selected to provide the coolant temperatures required to generate the desired supply air temperature. The required coolant temperature is determined by the performance characteristics of the cooling coils. A useful rule of thumb for preliminary analysis is that the supply air can be cooled to within 6°F (3°C) of the fluid temperature entering the coil. For example, for a 45°F (7°C) coil discharge air temperature, a fluid temperature of about 39°F (4°C) will be required.

Chilled-water storage systems storing water at 40°F (4°C) can produce supply air temperatures of 46°F to 50°F (8°C to 10°C). By using additives to lower the maximum density point and freezing point of water, stratified chilled-water storage systems may be able to store fluid at temperatures as low as 30°F (−1°C), allowing them to generate the same low air temperatures as ice systems.

Cold-air distribution systems with fluid supply temperatures of less than 40°F (4°C) may require increased insulation on pipes. Typically, the standard insulation thickness is increased by 0.5 in. (13 mm). Section 4.6.2 discusses determination of the appropriate insulation thickness to prevent condensation on ducts. The same concepts apply to pipes. Harmon and Yu (1993) discuss analytical considerations for preventing condensation on pipes carrying fluids of 20°F to 32°F (-6.7°C to 0°C).

4.2.1 ICE STORAGE

Coolant temperatures of 34°F to 39°F (1°C to 4°C) are typically available from ice storage systems. The available coolant temperature at a given time is dependent on the amount of ice remaining in storage and the rate at which the fluid must be cooled.

Ice storage systems must be sized with enough capacity to provide the required fluid temperature at the required flow rate for every hour of the design storage cycle. As ice is melted during the course of the discharge cycle, the effectiveness of heat transfer decreases and the minimum available fluid supply temperature rises. The minimum available supply temperature will also increase as the discharge rate increases.

Ice storage capacity is typically rated in terms of nominal storage capacity. The nominal capacity necessary to provide a given actual capacity depends on the required discharge temperature and on the rate at which storage must be discharged. The actual usable capacity is generally less than the nominal capacity. The lower the chilled-fluid supply temperature, the more additional nominal storage capacity must be provided.

Proper sizing of storage capacity for the required supply temperature and discharge rates is particularly important with cold-air distribution systems. Since cold-air systems use lower supply temperatures than systems with conventional distribution temperatures, they will be more likely to run out of stored cooling prematurely if they are sized incorrectly.

Ice storage should be configured in series with the chiller to ensure consistent low supply temperatures and to maintain a wide fluid temperature range. The two series options—chiller upstream and chiller downstream—are illustrated in Figure 4-1, with typical operating temperatures.

If the chiller is placed upstream of storage, it will operate at higher temperatures and will run more efficiently. However, to provide a fluid supply temperature of 36°F to 38°F (2°C to 3°C) leaving storage, more nominal storage capacity will be required than for a system operating with conventional distribution temperatures.

With a chiller-downstream arrangement, the chiller cools the fluid to the final supply temperature, and storage can be sized for a higher discharge temperature. This configuration can result in a significant reduction in the nominal storage capacity required. However, the chiller will operate at lower temperatures and lower efficiency. With chiller-downstream configurations, almost any temperature desired is achievable, but most systems are designed for 34°F to 38°F (1°C to 3°C) supply.

Some ice storage systems operate with different fluid supply temperatures, depending on whether cooling is supplied from ice or from a chiller. In such cases, the control sequence should allow for

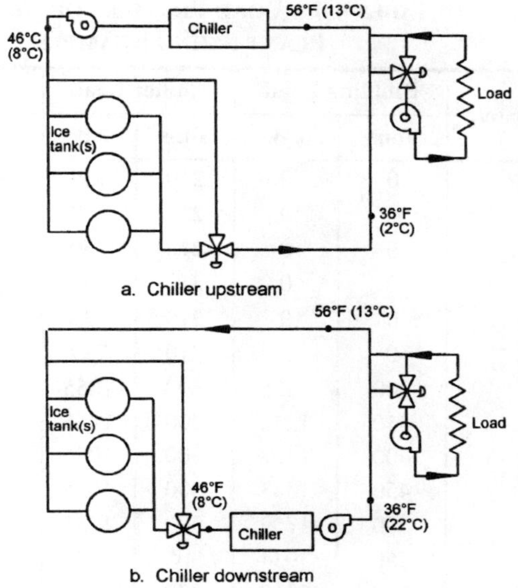

FIGURE 4-1 Storage plant configurations.

smooth ramping of the fluid supply temperature to avoid sudden changes in supply air temperature (Dorgan and Elleson 1987).

ASHRAE's *Design Guide for Cool Thermal Storage* (Dorgan and Elleson 1993) provides an extensive discussion of considerations for sizing and selecting ice storage equipment.

4.2.2 DESIGN EXAMPLE—ICE STORAGE COOLING PLANT SIZING

This example compares the required nominal storage capacity, chiller efficiency, and storage plant equipment cost for chiller-upstream and chiller-downstream configurations. The system is to serve a typical office building with a peak load of 625 tons (4,073 kW) and a total load of 6,100 ton-hours (77.2 GJ). The ice storage system is sized for load-leveling operation, with a 300-ton (1,055-kW) chiller and a storage requirement of 2,200 ton-hours (27.8 GJ). The load profile is summarized in Table 4-1.

The required chilled-fluid supply temperature while meeting load is 36°F (2°C), and the return fluid temperature is 56°F (13°C). The ice-making temperature is 24°F (-4°C). A maximum power rating of 0.7 kW/ton (5.0 COP) is specified for chilling. The chiller-

TABLE 4-1 LOAD PROFILE FOR STORAGE PLANT SIZING EXAMPLE

Hour	Building Load		Chiller Load		Storage Output	
	Tons	kW	Tons	kW	Tons	Kw
1	0	0	210	739	–210	–739
2	0	0	210	739	–210	–739
3	0	0	210	739	–210	–739
4	0	0	210	739	–210	–739
5	0	0	210	739	–210	–739
6	0	0	210	739	–210	–739
7	300	1055	300	1055	0	0
8	350	1231	300	1055	50	176
9	400	1407	300	1055	100	352
10	450	1583	300	1055	150	528
11	500	1759	300	1055	200	703
12	550	1934	300	1055	250	879
13	600	2110	300	1055	300	1055
14	625	2198	300	1055	325	1143
15	600	2110	300	1055	300	1055
16	550	1934	300	1055	250	879
17	475	1671	300	1055	175	615
18	400	1407	300	1055	100	352
19	300	1055	300	1055	0	0
20	0	0	210	739	–210	–739
21	0	0	210	739	–210	–739
22	0	0	210	739	–210	–739
23	0	0	210	739	–210	–739
24	0	0	210	739	–210	–739

upstream and chiller-downstream arrangements are illustrated in Figure 4-1. Storage requirements, in terms of nominal storage capacity required to meet the design load profile at the required temperature, are determined from manufacturers' performance and selection data. Chillers are also selected, using manufacturers' performance data, to provide ice making and chilling at the specified conditions. Storage and chiller costs are based on suppliers' budget estimates.

Table 4-2 summarizes the selection results. The chiller-downstream configuration is the lowest cost option because the reduction

TABLE 4-2 ICE STORAGE PLANT REQUIREMENTS

Description	Nominal Storage Capacity		Storage Cost $	Chiller Efficiency		Chiller Cost $	Chiller and Storage Cost $
	ton-hours	GJ		kW/Ton	COP		
Chiller downstream	2470	21.3	135,800	0.70	5.0	84,700	220,500
Chiller upstream, equal chiller efficiency	3420	43.3	188,100	0.70	5.0	71,300	259,400
Chiller upstream, equal chiller cost	3420	43.3	188,100	0.56	6.3	84,700	272,800

in nominal storage capacity offsets the increase in the chiller cost. Two chiller-upstream options are used to bracket the range of possible chiller selections for this configuration. The equal efficiency selection is a lower-priced chiller that operates at the same efficiency as the chiller-downstream option. The equal cost selection is the same machine as the chiller-downstream selection, operating at a higher efficiency.

Each project requires individual analysis of specific load requirements and equipment selections to determine the relative merits of the various options. However, for this example, the chiller-downstream arrangement is preferred on the basis of first cost. An economic analysis of operating costs would be required to evaluate the feasibility of the increased investment for the chiller-upstream option.

4.2.3 NONSTORAGE CHILLER

Chillers operating at 40°F to 42°F (4°C to 6°C) are commonly used with cold-air distribution systems supplying air at 47°F (8°C) or more. For lower supply air temperatures, chilled water may be cooled to as low as 36°F (2°C) without evaporator icing, although 38°F (3°C) is probably a practical minimum.

If lower chilled-fluid temperatures are required, freezing concerns can be eliminated by using a weak solution of glycol in the system. For the chilled-fluid temperatures normally used in cold-air distribution systems, a 10% concentration of ethylene or propylene glycol, with a freezing point of approximately 26°F (-3°C), provides ample protection against freezing.

4.2.4 DIRECT-EXPANSION SYSTEMS

Direct-expansion (DX) systems evaporate refrigerant directly in the cooling coil, rather than chilling an intermediate fluid. A conventional DX system uses a thermostatic expansion valve (TXV) to meter the amount of refrigerant fed to the cooling coil, so that all the refrigerant is evaporated and only superheated vapor leaves the coil. The advantages of DX systems for space conditioning include the following:

- The technology is widely used, well proven, and well known in the industry.
- The equipment is generally low in first cost.
- The equipment has low maintenance costs.
- The systems have a refrigerant charge of 3 to 5 pounds per ton (0.4 to 0.6 kg per kW) of capacity.

Direct-expansion systems also have some limitations. The discharge air temperature to prevent frosting or freezing of the coil generally must be above 45°F (7°C) unless special provisions are made. Control of discharge air temperature is less stable than with a chilled-fluid or liquid overfeed system. The coil has a relatively small thermal mass, and the discharge air temperature responds quickly to changes in compressor capacity.

Designers of cold-air distribution systems using DX refrigeration should observe the following recommendations:

Design for a minimum 45°F (7°C) discharge air temperature. As a general rule, a discharge temperature below 45°F (7°C) can lead to coil frosting and liquid carryover to compressors unless the design includes unusual provisions for compressor unloading and discharge temperature control.

For lower discharge air temperatures, provide for smooth capacity modulation. Use screw compressors with slide valve capacity control or open-drive reciprocating compressors with variable-speed drives. For compressors with multistep unloading, minimize the size of unloading steps and use electronic hot-gas bypass for fine discharge temperature control.

Consider a cascade-type system, as illustrated in Figure 4-2. The upstream coil could be a chilled-water coil, an indirect evaporative cooler, or a separate refrigeration circuit with its own compressor. This coil meets most of the variable load. The downstream coil sees

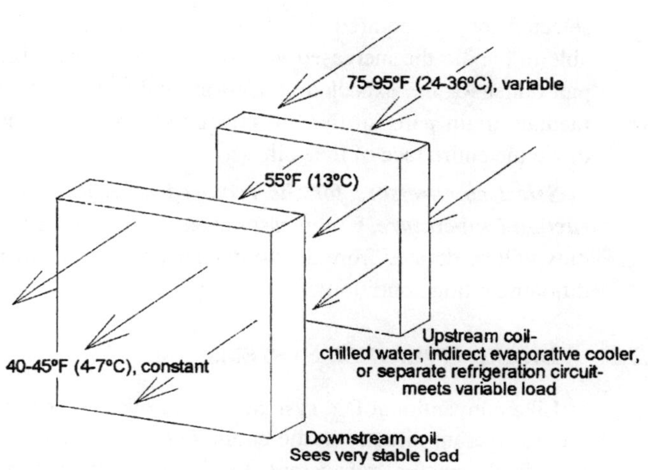

FIGURE 4-2 Cascaded coil arrangement for direct-expansion cold air distribution.

a stable load and can produce discharge temperatures as low as 40°F (4°C) with little difficulty.

Use multiple compressors with independent refrigeration circuits for improved part-load performance.

Provide positive subcooling of condensed liquid. This helps ensure a stable column of liquid behind the TXV so it doesn't have to pass two-phase flow.

Provide for stable head pressure control for a consistent pressure drop across the TXV. Provide floodback or variable-speed fan control on air-cooled condensers to minimize the rapid changes in head pressure caused by fan cycling.

Consider screw or scroll compressors. These compressors tend to be more liquid-tolerant than reciprocating compressors.

Design for oil return at low loads. Allow for sufficient refrigerant velocity or provide traps and double-suction risers. Final pipe sizing requires knowledge of the minimum refrigerant flow at the lowest compressor capacity. Hot gas bypass should be configured to introduce hot gas upstream of the coil. This arrangement helps prevent oil logging in the coil, since the mass flow of refrigerant through the coil is always equal to the flow through the compressor.

Select coils appropriate for the required supply air temperature. Especially if standard packaged rooftop air-handling units will be used, the coil provided with the unit should be specifically selected for the required heat transfer performance and should be able to handle the increased volume of condensate. Size the drain pan outlet for the expected condensate volume and provide intermediate drain pans so the full condensate volume does not flow down the entire face of the coil.

Select compressors for the required capacity at the reduced suction temperature. Compressor capacity for cold-air distribution duty will be derated from the nominal capacity at standard air-conditioning rating conditions.

4.2.5 LIQUID OVERFEED SYSTEMS

Like conventional DX systems, liquid overfeed systems evaporate refrigerant directly in the cooling coil. A liquid overfeed system feeds excess refrigerant to the coil, with a mixture of refrigerant vapor and liquid flowing from the coil to a low-pressure receiver. The vapor is drawn off to the compressor and the liquid is recirculated to the coil.

Liquid overfeed systems are more efficient than conventional DX systems because all internal surfaces of the liquid overfeed evaporator are wetted, resulting in better heat transfer performance. Discharge air temperature control is more stable because of the larger thermal mass of the coil and refrigerant. Oil return may also be improved, especially at low-load conditions.

There should be no difference in system life or reliability between liquid overfeed and conventional DX systems if both are properly designed, installed, commissioned, and maintained. However, liquid overfeed systems are more forgiving of errors. Compressors are protected from liquid slugs by the low-pressure receiver, which can lead to longer life and lower maintenance costs.

The installed cost of a liquid overfeed system is generally higher than for a conventional DX system. While overall maintenance costs can be lower, maintenance procedures may be unfamiliar to some operating personnel, and specific training may be required.

The refrigerant charge in a liquid overfeed system is much higher than in other types of DX systems. Designers contemplating the use of liquid overfeed should verify that the refrigerant charge does not exceed that allowed by *ANSI/ASHRAE 15-1994, Safety Code for Mechanical Refrigeration* (ASHRAE 1994a) or other applicable codes or standards.

Packaged air-handling units are not commonly available with liquid overfeed refrigeration systems, but such systems can be supplied by manufacturers of custom equipment.

Detailed information on liquid overfeed systems can be found in the *1994 ASHRAE Handbook—Refrigeration* (ASHRAE 1994b). Fields (1989) and Scofield and Fields (1989) discuss the use of liquid overfeed with cold-air distribution.

4.2.6 GLYCOL HEAT TRANSFER FLUIDS

Ethylene glycol or propylene glycol heat transfer fluids are often used in cooling plants for cold-air distribution systems. A solution of 25% by weight ethylene glycol in water is commonly used in ice storage applications. Lower concentrations are suitable for nonstorage chillers producing chilled-fluid temperatures of less than 38°F (3°C).

Designers of systems using glycol heat transfer fluid should take into account its higher viscosity, higher density, lower specific heat, and lower thermal conductivity compared to water. These differences result in increased flow requirements, reduced fluid-side heat transfer coefficients, and increased pressure drops. If glycol is to be used in an existing system designed for water, these disadvantages can generally be overcome by reducing the fluid supply temperature. A reduction of 2°F (1°C) will offset the air-side performance degradation resulting from the use of 25% ethylene glycol.

Performance ratings of all system components, including pumps, heat exchangers, cooling coils, chillers, etc., should be obtained from the respective manufacturers for the intended concentration. It is important that designers, installers, or operators do not change the concentration from the original design level without a thorough analysis of the effect of the change on system performance.

Only industrial-grade inhibited glycol formulated specifically for HVAC applications should be used in cooling systems. These mixtures contain corrosion inhibitors, reserve alkalinity, and anti-foam additives. Uninhibited glycols are more than twice as corrosive as tap water with most metals commonly used in HVAC piping systems. Automotive antifreezes can cause fouling and pump seal failures (Nussbaum 1990) and must never be used in HVAC applications.

Commercial glycol solutions are generally formulated for use in concentrations of 25% or more. At lower concentrations, the proportion of inhibitors may need to be increased. Designers should consult the fluid manufacturers for recommendations on increasing inhibitor levels at low glycol concentrations.

Glycol heat transfer fluid should be analyzed by a laboratory four times during the first year of operation and annually thereafter. The fluid should be treated according to the laboratory's and the manufacturer's recommendations for glycol concentration and proper levels of corrosion inhibitors and reserve alkalinity.

Gatley (1992) provides a detailed discussion of specific application considerations for systems using ethylene glycol. This publication includes the effects of the ethylene glycol solution on system flow rates, pressure drops, pump performance, and heat transfer, as well as additional information. Chapter 18 of the *1993 ASHRAE Handbook—Fundamentals* discusses properties of ethylene glycol and other heat transfer solutions. Chapter 12 of the *1996 ASHRAE Handbook—HVAC Systems and Equipment* (ASHRAE 1996) discusses the effects of glycol solutions on heat transfer, pump performance, and piping pressure drops. Nussbaum (1990) provides additional information on the application of glycol solutions.

4.3 SPACE LOAD CALCULATIONS

4.3.1 GENERAL CONSIDERATIONS

Space load calculations for cold-air distribution systems are, in most respects, the same as for conventional designs. This subject is covered thoroughly in chapter 26 of the *1993 ASHRAE Handbook—Fundamentals*, in the *Cooling and Heating Load Calculation Manual* (McQuiston and Spitler 1992), and in ASHRAE's *Air-Conditioning Systems Design Manual* (Lorsch et al. 1993).

There are some load calculation considerations that require special attention for cold-air distribution systems. The designer should consider all heat gains to supply air, including heat gains from supply fans and mixing box fans, and heat gains to ductwork. These sensible heat gains are considered part of the space load, since they increase the required supply air volume. Fan heat is discussed in detail in section 4.4 and duct heat gain is covered in section 4.5.

The low relative humidity maintained in spaces served by cold-air distribution systems may provide the same comfort level at an increased dry-bulb setpoint, compared to a conventional design at a higher relative humidity. Chapter 3 addresses this issue in more detail. This increase in the space setpoint, which may be 2°F (1°C) or more, can result in a decrease in space-cooling loads. However, in most cases, designers should size cold-air distribution systems to maintain the same dry-bulb temperature as a conventional design. This conservative approach provides a safety factor for designers and still allows for energy savings by operating at a higher space setpoint.

The space load calculation should also account for all latent loads. The following sections present considerations for calculation of space latent loads in cold-air distribution systems, and a design example.

4.3.2 SPACE LATENT LOADS

Accurate calculation of space latent loads is important for determining entering air conditions to the cooling coil and for evaluating the potential for condensation on diffusers and ductwork. Cold-air distribution systems typically maintain space relative humidity between 35% and 45% compared to the 50% to 60% generally present with conventional designs. The lower space humidity can result in a greater latent load from infiltration of moist air, and standard rule-of-thumb calculation methods developed for conventional designs may not apply.

Major sources of latent heat in the space include occupants, moisture-producing appliances, and infiltration of outside air. Latent loads from sources within the space are the same for cold-air distribution and conventional systems, since the moisture gains do not vary significantly with space relative humidity.

The latent load due to infiltration, however, does depend on the space humidity. The infiltration latent load is proportional to the difference between the humidity ratios of the outdoor air and the room air. This difference is higher for a space at 40% RH than for one at 50% RH, and designers of cold-air distribution systems should account for the resulting increased latent load.

Minimizing infiltration through good construction practices is particularly important for buildings using cold-air distribution. Engineers designing cold-air distribution systems should stress to owners and architects the importance of tight construction and of maintaining continuous air barriers and vapor retarders.

Many designers rely on the assumption that positive pressurization of the space will prevent infiltration. However, reliable control of space pressure is difficult to achieve. Space pressurization control depends on wind speed and direction, on the placement of outside and inside pressure sensors, and on accurate control of supply and return fan volumes. Most common strategies for maintaining positive space pressure do not do so in all zones under all conditions.

Moisture gain from infiltration during unoccupied periods when air-conditioning units are not running may also impose a significant latent load on the space. The water vapor from humid infiltration air will tend to be absorbed by hygroscopic materials in the space as the space humidity rises overnight. During the occupied period, when the space humidity is reduced, the moisture desorbs, causing a latent load.

Stewart et al. (1994, 1995) developed methods for determining the rates of vapor transfer through building wall materials. The researchers found that relatively large air pressure differentials are required to significantly reduce water vapor transmission under conditions typical of air-conditioned spaces in hot, humid climates.

In general, designers of cold-air distribution systems should not assume that infiltration will be prevented by positive space pressurization. Designers should include an allowance for infiltration in the space load calculation even if positive pressurization of the space is anticipated.

The amount of infiltration is calculated by the methods given in the "Infiltration and Ventilation" chapter of the *1993 ASHRAE Handbook—Fundamentals* and in ASHRAE's *Cooling and Heating Load Calculation Manual* (McQuiston and Spitler 1992).

Although the low space humidity present in cold-air distribution systems results in a greater driving force for diffusion of vapor through a building envelope, the vapor diffusion load is not a major concern. Even for extremely permeable construction, the magnitude of this load is so small that the increased diffusion has no significant effect on the air-conditioning load.

The following example illustrates the calculation of space latent loads for a space served by conventional and cold-air distribution designs.

4.3.3 DESIGN EXAMPLE—CALCULATION OF SPACE LATENT LOADS

Calculate the space latent loads for the example space, which comprises one floor of a multistory office building. Pertinent design data are as follows

Floor area:	16,650 ft^2	(1,550 m^2)
Wall area:	5,100 ft^2	(474 m^2)
Wall permeance:	2 perms	(0.115 $ng/s{\cdot}m^2{\cdot}Pa$)
Ceiling height:	10 ft	(3 m)
Infiltration:	0.5	air changes per hour (ACH)
Space sensible load:	350,000 Btu/h	(102,600 W)

Occupancy: per ASHRAE Standard 62-1989 for office space	143 ft^2/person (13.3 m^2/person) 117 occupants	
Outside air dry bulb	91°F	(33°C)
Outside air wet bulb	74°F	(23°C)
Room temperature:	75°F	(24°C)
Room humidity:	50% RH, for conventional design 40% RH, for cold-air distribution design	

Values of vapor pressure (*vp*) and humidity ratio (*W*) for the air conditions in question are given in Table 4-3.

TABLE 4-3 VAPOR PRESSURE AND HUMIDITY RATIO CONDITIONS FOR LATENT LOAD EXAMPLES

Air Condition	*vp*		*W*	
	in. Hg	kPa	lb_w/lb_a	kg_w/kg_a
91°F db/74°F wb: (33°C db/23°C wb)	0.6607	2.2312	0.0141	0.0141
75°F db/50% rh: (24°C db/50°C rh)	0.4378	1.4786	0.0093	0.0093
75°F db/40% rh: (24°C db/40% rh)	0.3503	1.1829	0.0073	0.0073

Infiltration Load

The latent cooling load due to infiltration into the space is calculated by

$$q_{\text{lat}} = C_2 Q_{\text{inf}}(W_{\text{OA}} - W_{\text{RM}}) \qquad (4\text{-}4)$$

where

q_{lat} = latent load due to infiltration, Btu/h (W)

Q_{inf} = infiltration volume flow, cfm (L/s)

W_{OA} = outside air humidity ratio, lb water vapor/lb dry air (kg/kg)

W_{RM} = space humidity ratio, lb_w/lba (kg/kg)

C_2 = 4,840 Btu·min·$lb_a/(h{\cdot}ft^3{\cdot}lb_w)$(3,002 W·s/L)

The infiltration flow rate for this example, with a typical rate of 0.5 ACH, is 1,388 cfm (655 L/s). For the conventional design, the latent cooling load imposed by this infiltration is

$$\begin{aligned} Q_{lat} &= (4,840)(1,388)(0.0141 - 0.0093) \\ &= 32,246 \text{ Btu/h} \end{aligned}$$

$$\begin{pmatrix} Q_{lat} = (3002)(655)(0.0141 - 0.0093) \\ = 9,438 \text{ W} \end{pmatrix}$$

This load is equivalent to 1.94 Btu/h·ft² (6.09 W/m²). For the cold-air distribution design, a similar calculation leads to a latent infiltration load of 45,68² Btu/h (13,371 W), or 2.74 Btu/h·ft² (8.63 W/m²). This is an increase of 42% for the cold-air system, and it represents a small but significant impact on the total space load.

Occupant Latent Load

The latent load due to occupants in the space can be determined from tabulated values, such as those given in McQuiston and Spitler (1992) and in chapter 26 of the *1993 ASHRAE Handbook—Fundamentals*. (Note that the latent heat values in the table printed in the *1993 Fundamentals* are incorrect. The correct values are given in McQuiston and Spitler (1992), and in Appendix A, "Additions and Corrections," of the *1994 ASHRAE Handbook—Refrigeration*.

For occupants performing moderately active office work, the recommended rate of latent heat gain is 200 Btu/h (58.6 W) per person. For the example system, this amounts to a space latent load of 23,400 Btu/h (6858 W), or 1.41 Btu/h·ft² (4.42 W/m²). This value is the same for both the conventional and cold-air designs.

Vapor Diffusion Load

The amount of water vapor diffusing through the exterior walls is given by

$$M_V = MA\frac{\Delta p}{C_3} \tag{4-5}$$

where

M_v = mass of water vapor, lb/h (g/s)

M = permeance, gr/h·ft²·in. Hg (ng/s·m²·Pa)

A = wall area, ft² (m²)

Δp = vapor pressure differential, in. Hg (kPa)

C_3 = 7,000 gr/lb$_w$ (1 kg/kg)

For the conventional design, the total diffusion is

$$M_v = \frac{(2)(5100)(0.6607 - 0.4378)}{7000} = 0.325 \text{lb}_w/\text{hr} \qquad \begin{pmatrix} M_v = (.115)(474)(2.2312 - 1.4786) \\ = 41.02 \text{ g/s} \end{pmatrix}$$

The resulting space latent load is calculated by multiplying this diffusion rate by the vaporization enthalpy difference of 1,076

Btu/lb_w (2.502 kJ/kg). The latent load is 349 Btu/h (103 W), or 0.068 Btu/h·ft^2 (0.22 W/m^2). For the cold-air distribution design, the same calculation results in a load of 487 Btu/h (143 W), or 0.095 Btu/h·ft^2 (0.30 W/m^2).

Most walls have an overall permeance less than the 2 perms (0.115 ng/s·m^2·Pa) selected for this example. Even with this worst-case assumption, the calculated diffusion loads are very low. These small values show that the latent load due to vapor diffusion can be neglected in cold-air distribution design.

Summary of Space Latent Loads

Table 4-4 summarizes the space latent loads calculated for this example.

TABLE 4-4 EXAMPLE SPACE LATENT LOAD SUMMARY

Space Conditions	75°F (24°C) db/50%rh				75°F (24°C) db/40%rh			
	Btu/h	W	Btu/h·ft^2	W/m^2	Btu/h	W	Btu/h·ft^2	W/m^2
Vapor Diffusion	349	103	0.068	0.22	487	143	0.095	0.30
Infiltration	32,246	9,438	1.94	6.09	45,682	13,371	2.74	8.63
Occupants	23,400	6,858	1.41	4.42	23,400	6,858	1.41	4.42
Total Latent load	55,995	16,399	3.42	10.75	69,569	20,372	4.25	13.35

Table 4-4 shows that the space latent load for the cold-air distribution system is higher than that for the conventional design because of the additional energy expended to dehumidify infiltration air to the lower space humidity condition. This increased latent load increases the total space load and decreases the space sensible heat ratio, as shown in Table 4-5.

TABLE 4-5 EXAMPLE SPACE LOADS AND SENSIBLE HEAT RATIOS

Space Condition	75°F db/50%rh (24°C db/50% rh)		75°F db/40%rh (24°C db/40% rh)	
	Btu/h	W	Btu/h	W
Space Sensible Load	350,000	102,585	350,000	102,585
Space Latent Load	55,995	16,399	69,569	20,372
Total Space Load	405,995	118,984	419,569	122,957
Space Sensible Heat Ratio	0.86		0.83	

4.4 FAN HEAT GAIN AND TEMPERATURE RISE

4.4.1 FAN HEAT

Fan heat is a significant cooling load that must be included in the cooling load calculation. Supply fans produce the largest heat gains, but heat from return fans and mixing-box fans must also be considered if these are present in the system.

If the supply fan is downstream of the cooling coil in a draw-through configuration, the fan heat increases the supply air temperature, increasing the supply air volume required to meet a given space load. If the fan is upstream of the coil in a blow-through configuration, the fan heat is absorbed directly by the coil. In either case, the fan heat adds to the cooling coil load.

Fan heat gain can be calculated from the following equations:

If the fan motor is in the airstream.

$$Q_{fan} = C_4 \frac{Qptot}{\eta_{fan}\eta_{mot}} \tag{4-6}$$

or

$$q_{fan} = C_5 \frac{bhp}{\eta_{mot}} \tag{4-7}$$

If the fan motor is outside the airstream:

$$Q_{fan} = C_4 \frac{Qptot}{\eta_{fan}} \tag{4-8}$$

or

$$q_{fan} = C_5 bhp \tag{4-9}$$

where

q_{fan} = fan heat gain, Btu/h (W)
Q = fan airflow volume, cfm (L/s)
$ptot$ = fan total pressure, in. w.g. (Pa)
η_{fan} = fan efficiency
η_{mot} = motor and drive efficiency
bhp = fan brake horsepower (kW)
C_4 = 0.40 Btu·min/in. w.g.·ft^3 (9.99×10^{-4} W·s/L·Pa)
C_5 = 2,545 Btu/hp·h (1 kW/kW)

If the specific horsepower rating for the fan to be used is not known, Equation 4-6 or 4-8 can be used with an estimate of the fan efficiency.

Fan efficiency for air-handling unit supply and return fans is generally between 50% and 70%, with a typical average of about 65%.

Motor efficiency is typically 80% to 90%. Typical motor efficiencies for a range of motor sizes are tabulated in chapter 26 of the *1993 ASHRAE Handbook—Fundamentals* and in ASHRAE's *Cooling and Heating Load Calculation Manual* (McQuiston and Spitler 1992). Premium efficiency motors may have efficiencies several points higher than the standard motors tabulated by these sources.

Belt-drive losses are commonly estimated as 3% of the motor power rating. The overall motor and drive efficiency in Equations 4-6 and 4-7 is the motor efficiency times 0.97. Losses associated with variable-frequency drives (VFDs), which may be 5% to 6% of the motor power rating, normally occur at the VFD cabinet and do not add heat to the airstream.

If the fan horsepower rating is known, the fan heat gain can be calculated from Equation 4-7 or 4-9. Note that the actual fan horsepower is used in these equations, rather than the motor horsepower. The fan horsepower is commonly available from the manufacturer's selection data.

4.4.2 FAN TEMPERATURE RISE

The fan temperature rise must be known in order to calculate the actual supply air temperature for draw-through fans and to calculate the cooling coil entering air condition for blow-through fans. The fan temperature rise can be calculated from the following equations:

$$\Delta T_{fan} = C_6 \frac{P_{tot}}{\eta} \qquad (4\text{-}10)$$

where

C_6 = 0.364°F/in. w.g. (0.00083°C/Pa)

$\eta = \eta_{fan}$ if the fan is outside the airstream and

$\eta = \eta_{fan}\eta_{mot}$ if the fan is in the airstream

$$\text{or } \Delta T_{fan} = \frac{q_{fan}}{C_1 Q} \qquad (4\text{-}11)$$

where

C_1 = 1.1 Btu·min/ft^3·h·°F (1.24 W·s/m^3·°C)

The fan efficiency can be determined from the fan manufacturer's data and the following formula:

$$\eta_{fan} = Q\frac{P_{tot}}{C_7 bhp} \tag{4-12}$$

where

C_7 = 6354 in. w.g.·ft³/hp·min (1,000 Pa·L/s·W)

4.4.3 FAN-POWERED MIXING BOXES

Where fan-powered mixing boxes temper primary supply air with recirculated room air before introducing it to the space, additional heat gains must be considered. The mixing box fans add heat to the supply air, as does the recirculation of plenum air that has absorbed some plenum heat.

Fan power ratings for fan-powered mixing boxes are typically listed in terms of watts for a given model and operating condition. The heat gain is determined directly from the power input as:

$$q_{\text{fan}} = C_8 P \tag{4-13}$$

where

P = fan power rating, W (W)

C_8 = 3.413 Btu/W·h (1 W/W)

The heat gain can also be calculated using Equation 4-6. Note that the overall efficiency of mixing-box fans is quite low. The efficiency of fractional horsepower motors in the sizes typically used in these units is about 35%. The typical overall efficiency, calculated from a survey of rating data for several manufacturers' products, is in the range of 5% to 10%.

The temperature rise is calculated from the fan heat gain by Equation 4-11.

An additional heat gain arises from the recirculation of plenum air that has absorbed some of the plenum heat load. Plenum heat loads normally include a fraction of the lighting load, as well as a portion of the roof load if the return air plenum is below a roof. These loads are normally not part of the space load and do not affect the required supply air volume. However, when fan-powered mixing boxes recirculate some of the plenum air to the space, the primary air quantity must be increased to offset the return air heat gain.

4.4.4 DESIGN EXAMPLES—FAN HEAT GAIN AND TEMPERATURE RISE

1. Conventional System Supply Fan

Calculate the estimated fan heat gain and temperature rise for a supply fan to be selected for the following duty:

Fan volume:	20,400 cfm (9,620 L/s)
Fan total pressure:	3.0 in. w.g. (746 Pa)

Fan and motor are in the airstream

Since the actual equipment is not yet selected, the fan efficiency is estimated to be 60% and the motor efficiency is estimated to be 88%. From Equation 4-6, the heat gain is

$$q_{fan} = (0.40)(20,400)\frac{3.0}{(0.60)(0.88)(0.97)}$$

$$= 47,798 \text{ Btu/h}$$

$$\left(q_{fan} = (9.99 \times 10^{-4})(9620)\frac{746}{(0.60)(0.88)(0.97)} = 13998 \text{ W}\right)$$

From Equation 4-10, the temperature rise is

$$\Delta T_{fan} = 0.364\frac{3.0}{(0.60)(0.88)(0.97)}$$

$$= 2.1°\text{F}$$

$$\Delta T_{fan} = \left(0.00083\frac{746}{(0.60)(0.88)(0.97)}\right)$$

$$= 1.2°\text{C}$$

If this fan is to be configured in a blow-through arrangement, the fan heat is added directly to the coil load and the dry-bulb temperature entering the coil is increased by 2.1°F (1.2°C). For a draw-through arrangement, the fan heat is absorbed by the supply air and the supply air temperature is increased by 2.1°F (1.2°C).

2. Cold-Air Distribution System Supply Fan

Calculate the estimated fan heat gain and temperature rise for a supply fan with the same parameters as number 1 above but with a total volume of 12,830 cfm (5,995 L/s).

From Equation 4-6, the heat gain is

$$q_{fan} = (0.40)(12{,}830)\frac{3.0}{(0.60)(0.88)(0.97)}$$
$$= 30{,}061 \text{ Btu/h}$$

$$\left(q_{fan} = (9.99 \times 10^{-4})(5995)\frac{746}{(0.60)(0.88)(0.97)} = 8723 \text{ W}\right)$$

From Equation 4-10, the temperature rise is

$$\Delta T_{fan} = 0.364\frac{3.0}{(0.60)(0.88)(0.97)} = 2.1°\text{F}$$

$$\Delta T_{fan} = 0.00083\frac{746}{(0.60)(0.88)(0.97)} = 1.2°\text{C}$$

The heat gain for the downsized supply fan is reduced in direct proportion to the reduction in supply air volume, but the temperature rise remains the same.

3. Fan-Powered Mixing Box

Determine the heat gain and primary air requirement for the following zone supplied by a parallel fan-powered mixing box:

Sensible load:	25,000 Btu/h (7,328 W)
Plenum load:	5,000 Btu/h (1,465 W)
Primary air temperature entering box:	48°F (9°C)
Space temperature:	75°F (24°C)
Final supply air temperature:	55°F (13°C)
Primary air volume to meet space load:	842 cfm (397 L/s)
Recirculated air volume:	290 cfm (137 L/s)
Temperature rise to recirculated air:	4°F (2.2°C)
Fan power rating:	110 W

The additional space load due to the temperature rise of the recirculated air is calculated as follows:

$$q_{recirc} = C_1 Q_{recirc} \Delta T_{recirc} \qquad (4\text{-}14)$$

where

q_{recirc} = additional space load due to temperature rise of the recirculated air

Q_{recirc} = recirculated air volume
ΔT_{recirc} = temperature rise to recirculated air
C_1 = 1.1 Btu·min/ft^3·h·°F (1.24 W·s/m^3·°C)

$$q_{recirc} = (1.1)(290)(4) = 1276 \text{ Btu/h} \qquad \left(q_{recirc} = (1.24)(137)(2.2) = 374 \text{ W} \right)$$

The additional primary airflow required to offset this load is

$$Q_{primry,recirc} = \frac{q_{recirc}}{C_1(T_m - T_{primary})} = \frac{1276}{1.1(75-48)} = 43 \text{ cfm} \qquad \left(= \frac{376}{1.24(24-9)} = 20 \text{ L/s} \right) \tag{4-15}$$

From Equation 4-13, the fan heat gain is

$$q_{fan} = 3.413P = (3.413)(110) = 375 \text{ Btu/h} \qquad \left(q_{fan} = 1.0P = (1.0)(110) = 110 \text{ W} \right)$$

The additional primary airflow required to offset the fan heat is

$$Q_{primry,recirc} = \frac{q_{fan}}{C_1(T_m - T_{primary})} = \frac{375}{1.1(75-48)} = 13 \text{ cfm} \qquad \left(= \frac{110}{1.24(24-9)} = 6 \text{ L/s} \right) \tag{4-16}$$

The mixing-box heat gains result in an additional primary air requirement of 56 cfm (26 L/s), an increase of about 7%. Note that these increases are directly dependent on the return air temperature rise. As discussed in section 4.5, heat gains to supply ducts can cause a cooling effect on plenum return air. However, these effects are not uniform throughout the distribution system. As a safety factor, designers may wish to neglect these cooling effects when calculating primary air requirements for fan-powered mixing boxes.

4.5 DUCT HEAT GAIN

4.5.1 GENERAL CONSIDERATIONS

Heat gain to supply air through ductwork typically results in a temperature rise of 3°F to 5°F (1.6°C to 2.7°C). In a system where air is returned to the air-handling unit via the ceiling plenum, heat gain to supply ducts is normally accompanied by a corresponding heat loss from return air, and the cooling effect is not entirely lost to the system. However, the duct heat gain is added to the space-cooling load, increasing the required supply air volume.

Duct heat gain depends on the temperature difference between the air inside and the air surrounding the duct, on the duct surface area, and on the overall heat transfer coefficient, or U-factor.

The temperature difference across the duct wall may be 40% to 70% greater in a cold-air distribution system than in a conventional design, while the surface area is typically 15% to 40% less. With equal insulation levels, the heat gain for a given cold-air system might be from 15% less than to 40% greater than a conventional design. Normally, increased insulation thickness is specified for cold-air distribution systems, reducing the heat gain to 40% to 80% of that in conventional systems. However, even with this lower heat gain, the supply air temperature rise in the cold-air distribution system will be equivalent to or somewhat greater than that in the conventional design, because of the reduced supply air quantity.

The duct heat transfer coefficient depends on the type, density, and thickness of insulation. For design calculations, the overall duct heat transfer coefficient can be approximated by the heat transfer coefficient of the insulation plus the external air film, neglecting the thermal resistance of the duct wall.

$$U_{duct} = \frac{1}{R_i + R_s} \qquad (4\text{-}17)$$

where

U_{duct} = overall duct heat transfer coefficient, Btu/h·ft²·°F (W/m²·K)

R_i = insulation thermal resistance, °F·ft²·h/Btu (m²·K/W)

R_s = duct surface air film thermal resistance, 0.62 °F·ft²·h/Btu (0.109 m²·K/W)

and

$$R_i = \frac{t_i}{k_i} \tag{4-18}$$

where

t_i = insulation thickness, in. (mm)

k_i = insulation thermal conductivity, Btu·in./h·ft²·°F (W/m²·K)

TABLE 4-6 TYPICAL INSULATION VALUES FOR GLASS FIBER DUCT WRAP

I-P Units

Nominal Density	Nominal Thickness	Out-of Package R-value	Installed Thickness	Installed R-value	Surface R-value	Overall U-value
lb/ft³	in.	°F·ft²·h/Btu	in	°F·ft²·h/Btu	°F·ft²·h/Btu	Btu/h·°F·ft²
0.75	1.5	5.0	1.125	4.2	0.62	0.2075
	2	6.7	1.5	5.6	0.62	0.1608
	3	10.0	2.25	8.3	0.62	0.1121
1.0	1	3.7	0.75	3.0	0.62	0.2762
	1.5	5.6	1.125	4.5	0.62	0.1953
	2	7.4	1.5	6.0	0.62	0.1511
1.5	1.5	6.0	1.125	4.8	0.62	0.1845
	2	8.0	1.5	6.4	0.62	0.1425

SI Units

Nominal Density	Nominal Thickness	Out-of Package R value	Installed Thickness	Installed R-value	Surface R-value	Overall U-value
Kg/m³	mm	K·m²/W	mm	K·m²/W	K·m²/W	W/m²·K
12	38.1	0.8805	28.6	0.7396	0.109	1.1782
	50.8	1.1798	38.1	0.9861	0.109	0.9130
	76.2	1.7611	57.2	1.4616	0.109	0.6365
16	25.4	0.6561	12.0	0.5283	0.109	1.5682
	38.1	0.9861	28.6	0.7924	0.109	1.1089
	50.8	1.3031	38.1	1.0566	0.109	0.8579
24	38.1	1.0566	28.6	0.8453	0.109	1.0475
	50.8	1.4088	38.1	1.1270	0.109	0.8091

If there are multiple layers of insulation, the insulation thermal resistance is equal to the sum of the R-values of the individual layers.

Insulation thermal resistance or conductivity values are generally published by insulation manufacturers for their products. Table 4-6 lists typical insulation values for glass fiber duct wrap.

Typical thermal conductivities are also tabulated in the *Cooling and Heating Load Calculation Manual* (McQuiston and Spitler 1992, Table 4.2) and in chapter 22, Table 10, of the *1993 ASHRAE Handbook—Fundamentals*. Typical conductivities for glass fiber blanket insulation at a mean temperature of 50°F (10°C) range from 0.29 to 0.22 Btu·in./h·ft^2·°F (0.04 to 0.003 W/m·K) for insulation densities from 0.75 to 3.0 lb/ft^3 (12 to 48 kg/m^3).

External blanket insulation is normally compressed somewhat during installation, and the conductivity increases above the uncompressed value. Designers should ensure that "as-installed" values, taking compression into account, are used for design calculations. These values are typically provided by insulation manufacturers.

For uninsulated ducts, internally lined ducts, and flexible ducts with air-pervious liners, the heat transfer coefficient increases with increasing duct air velocity. The *1993 ASHRAE Handbook—Fundamentals* presents data showing variations in duct heat transfer coefficients with air velocity for various insulation types (ASHRAE 1993, p. 32.14).

4.5.2 DUCT HEAT GAIN CALCULATIONS

Duct heat gain and temperature rise can be estimated using the following formulas, adapted from the *1993 ASHRAE Handbook—Fundamentals* and from Dorgan and Elleson (1988). A more exact method of calculating duct heat gain is presented in McQuiston and Spitler (1992, p. D.10).

$$q_{duct} = \frac{UPL_{duct}}{C_9}\left(T_a - \frac{T_e - T_l}{2}\right) \tag{4-19}$$

$$T_l = \frac{T_e(y-1) + 2T_a}{y+1} \tag{4-20}$$

where

$$y = C_{10} A_{cs} V \frac{\rho}{UPL_{duct}} \quad \text{for rectangular ducts} \tag{4-21}$$

$$y = C_{11} D V \frac{\rho}{UPL_{duct}} \quad \text{for round ducts} \tag{4-22}$$

and

A_{cs} = cross-sectional area of duct, in.^{2} (mm^2)
V = average velocity, fpm (m/s)
D = diameter of duct, in. (mm)
L_{duct} = duct length, ft (m)
Q_{duct} = heat gain through duct walls, Btu/h (W)
U = overall heat transfer coefficient of duct wall, Btu/h·ft^2·°F (W/m^2·K)
P = outside perimeter of bare or insulated duct, in. (mm)
ρ = air density, lb/ft^3 (kg/m^3)
T_e = temperature of air entering duct section, °F (°C)
T_l = temperature of air leaving duct section, °F (°C)
T_a = temperature of air surrounding duct, °F (°C)
C_9 = 12 in./ft (1,000 mm/m)
C_{10} = 2.4 Btu·ft·min/lb·in.·h (2.01 W·m·s/mm·kg)
C_{11} = 0.6 Btu·ft·min/lb·in.·h (0.5 W·m·s/mm·kg)

To determine heat gain and temperature rise for a given duct run, these equations are solved for each duct section, with the leaving temperature for one section used as the entering temperature for the next. This type of repetitive calculation is easily performed using a computer program.

The temperature rise is normally calculated for the longest duct run in the system, and the resulting final supply air temperature is used for determining zone supply air requirements. Some designers use a standard allowance of 3°F (1.6°C) for duct temperature rise in typical office occupancies. For unusually long runs or where ducts are not insulated to the recommended levels, the actual rise should be calculated using Equation 4-20.

The total duct heat gain for the entire system can be calculated by applying Equation 4-19 to each duct section. Designers can obtain a less precise estimate by assuming that the temperature rise calculated for the longest duct run applies to the entire system airflow.

$$Q_{duct,total} \cong C_1 Q_{sa} \Delta T_{duct} \tag{4-23}$$

where

$q_{duct,total}$ = total duct system heat gain

Q_{sa} = supply air volume

ΔT_{duct} = duct temperature rise

C_1 = 1.1 Btu·min/ft^3·h·°F (1.24 W·s/m^3·°C)

This approach provides a generous estimate of the total heat gain, since some of the supply air will experience less than the maximum temperature rise.

The supply air temperature rise may increase to as much as 5°F to 10°F (3°C to 6°C) as the supply airflow is reduced in VAV systems at partial loads. The amount of heat transferred across the duct wall is essentially constant, but with a smaller air volume to absorb the heat, the temperature rise is greater.

An increased temperature rise at low loads can help offset any decline in diffuser performance caused by the reduced diffuser velocity. However, if one zone will have high loads even when the load on the remainder of the system is low, designers should determine airflow requirements for that zone based on the higher duct temperature rise.

4.5.3 DESIGN EXAMPLE—CALCULATION OF DUCT HEAT GAIN AND TEMPERATURE RISE

Evaluate the duct heat gain and temperature rise for the cold-air distribution duct run illustrated in Figure 4-3, supplying nominal 45°F (7°C) supply air, with a 78°F (25°C) ambient temperature, with 1-, 1.5-, and 2-in. (25-, 38-, and 51-mm) external insulation. Calculate heat gain and temperature rise for comparable duct runs, resized for nominal 50°F (10°C) and 55°F (13°C) supply air, with the same insulation levels.

Table 4-7 illustrates the calculation of duct heat gain and temperature rise for this example. This calculation can easily be implemented using a computer spreadsheet, with the following inputs:

Length (L_{duct}), width (W), and height (H) of each duct section, in. (mm)

Volume (Q) of airflow through each duct section, cfm (L/s)

Air density (ρ) at the temperature entering the first duct section, lb/ft^3 (kg/m^3)

Insulation thickness (t_i), in. (mm)

Overall heat transfer coefficient (U) of the duct, Btu/h·ft^2·°F (W/m^2·K)

Ambient temperature (T_a) of the air surrounding the duct, °F (°C)

Entering temperature (T_e) of the air into the first section of the duct, °F (°C)

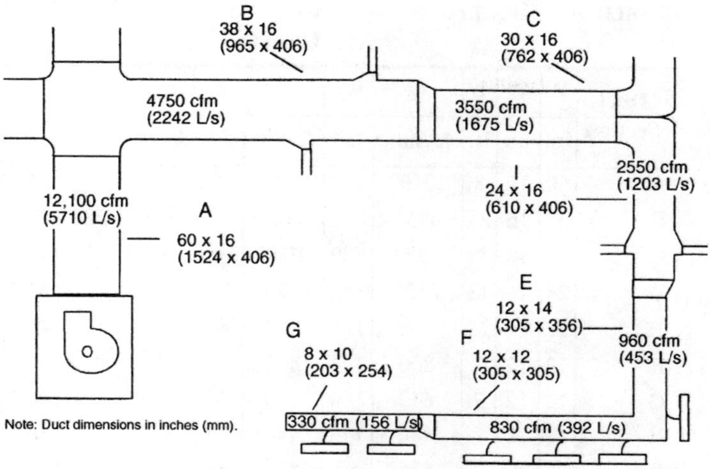

FIGURE 4-3 Duct run for heat gain example. Duct size and airflows are for 45°F supply air system.

The overall heat transfer coefficient is approximated from the thermal resistances of the duct insulation and the surface air film by Equation 4-17.

The spreadsheet uses the following formulas to calculate intermediate quantities and to determine the temperature rise and heat gain for each section:

$$A_{cs} = WH \tag{4-24}$$

$$V = Q\frac{C_{12}}{A_{cs}} \tag{4-25}$$

where

$$C_{12} = 144 \text{ in.}^2/\text{ft}^2 \ (1000 \text{ mm}^2\cdot\text{m/L})$$

The perimeter calculation is based on the mean of the inside and outside duct dimensions:

$$\begin{aligned} P &= 2[(W + t_i) + (H + t_i)] \\ &= 2(W + H + 2t_i) \end{aligned} \tag{4-26}$$

The leaving temperature for each section is calculated from Equations 4-20 and 4-21. The leaving temperature for one section is the entering temperature for the next. The total tempera-

TABLE 4-7 DUCT HEAT GAIN CALCULATIONS FOR 45°F (7°C) SYSTEM (I-P UNITS)

Duct Section	W	H	L	Q	A_{cs}	V	P	q_{duct}	Y	T_e	T_l
	in.	in.	ft.	ft³/min	in²	ft/min	in	Btu/h		°F	°F
A	60	16	8	12100	960	1815	160	563	1607	45.0	45.0
B	38	16	26	4750	608	1125	116	1251	283	45.0	45.3
C	30	16	42	3550	480	1065	100	1707	153	45.3	45.7
D	24	16	13	2550	384	956	88	466	400	45.7	45.9
E	12	14	25	960	168	823	60	611	113	45.9	46.4
F	12	12	7	800	144	800	56	155	366	46.4	46.6
G	12	12	7	640	144	640	56	154	293	46.6	46.8
H	12	12	7	480	144	480	56	153	220	46.8	47.1
I	8	10	7	576	80	527	44	119	186	47.1	47.4
J	8	10	7	288	80	288	44	117	93	47.4	48.1

Total temperature rise = 3.1°F.

DUCT HEAT CALCULATIONS FOR 45°F (7°C) SYSTEM (SI UNITS)

Duct Section	W	H	L	Q	A_{cs}	V	P	q_{duct}	Y	T_e	T_l
	mm	mm	m	L/s	mm²	m/s	mm	W		°C	°C
A	1520	410	2.4	5710	623200	9.2	4063	154	1617	7.2	7.2
B	970	410	7.9	2241	397700	5.6	2963	364	268	7.2	7.4
C	760	410	12.8	1675	315700	5.3	2563	503	143	7.4	7.6
D	610	410	3.9	1203	250100	4.8	2243	135	379	7.6	7.7
E	310	360	7.6	453	111600	4.1	1543	176	108	7.7	8.1
F	310	300	2.1	377	93000	4.1	1423	45	348	8.1	8.2
G	310	300	2.1	302	93000	3.2	1423	45	278	8.2	8.3
H	310	300	2.1	226	93000	2.4	1423	44	208	8.3	8.4
I	200	250	2.1	151	50000	3.0	1103	34	179	8.4	8.6
J	200	250	2.1	75	50000	1.5	1103	43	90	8.6	9.0

Total temperature rise = 1.8°C.

ture rise is the final leaving temperature minus the initial entering temperature.

Table 4-8 summarizes the results of similar calculations for comparable duct runs sized for nominal 45°F, 50°F, and 55°F (7.2°C, 10°C, and 12.8°C) supply air, with insulation thicknesses of 1-, 1.5-, and 2-in. (25-, 38-, and 50-mm) external insulation.

Additional examples of duct heat gain under various conditions are given in Dorgan and Dorgan (1995).

TABLE 4-8 DUCT TEMPERATURE RISE SUMMARY

Supply Air Temperature		45°F (7.2°C)		50°F (10.0°)		55°F(12.8°C)	
Insulation Thickness		ΔT		ΔT		ΔT	
in.	mm	°F	°C	°F	°C	°F	°C
1"	25	4.4	2.4	3.3	1.8	2.3	1.3
1.5"	38	3.8	2.1	2.8	1.6	1.9	1.1
2"	50	3.1	1.7	2.3	1.3	1.6	0.9

Notes: Shaded boxes indicate typical insulation levels for each supply temperature. Duct systems were resized for each supply temperature.

4.6 DUCT INSULATION REQUIREMENTS

4.6.1 GENERAL REQUIREMENTS

Insulation on ductwork and other air distribution equipment serves to reduce heat gain to the supply air and to prevent condensation of ambient moisture. Insulation requirements are increased in cold-air distribution systems because of the lower supply air temperature. The higher temperature difference to ambient air increases the driving force for heat gain, and there is an increased chance of some surfaces being cooled below the ambient dew point and causing condensation. For most applications, insulation specified for the purpose of limiting heat gain will also be sufficient to prevent condensation.

To minimize heat gain and the resulting increase in supply air volume, cold-air distribution systems should be designed with thicker insulation than that used in standard designs. Duct heat gain is calculated by the methods given in section 4.5. Designers should evaluate the benefits of reduced heat gain and supply air volume with added insulation vs. the costs for each system. However, in general, where 1-in. (25-mm) external blanket insulation is used with conventional supply temperatures, 2-in. (51-mm) insulation is recommended for systems with nominal supply air temperatures of 40°F to 45°F (4°C to 7°C), and 1.5-in. (38-mm) thickness is recommended for nominal supply temperatures of 46°F to 50°F (8°C to 10°C).

Good workmanship in the application of duct insulation is essential for cold-air distribution systems. Complete coverage of ducts and air distribution equipment and a continuous, intact vapor retarder are necessary to prevent excessive heat gains and unwanted condensation.

4.6.2 INSULATION THICKNESS TO PREVENT CONDENSATION

To prevent condensation on ducts, the insulation must meet the following conditions:

- It must be thick enough to maintain the duct surface temperature higher than the ambient dew-point temperature.
- It must cover all surfaces that may be cooled below the ambient dew point.
- It must have a complete, effective vapor retarder to prevent moisture from diffusing into and condensing inside the insulation.

The insulation thickness required to prevent condensation for a given ambient condition is determined by a heat transfer analysis of the duct and insulation.

The surface temperature can be calculated from:

$$T_s = T_{sa} + \left[(T_a - T_{sa})\frac{R_i}{R_i + R_s}\right] \qquad (4\text{-}27)$$

For the surface temperature to be equal to or greater than the ambient dew point, the minimum insulation thermal resistance is

$$R_i = R_s \frac{T_{dp} - T_{sa}}{T_a - T_{dp}} \qquad (4\text{-}28)$$

and the corresponding insulation thickness is

$$t_i = k_i R_i \qquad (4\text{-}29)$$

where

T_s = duct surface temperature, °F (°C)

T_{sa} = supply air temperature, °F (°C)

T_a = ambient dry-bulb temperature surrounding duct, °F (°C)

T_{dp} = ambient dew-point temperature surrounding duct, °F (°C)

R_i = insulation thermal resistance, °F·ft^2·h/Btu (m^2·K/W)

R_s = surface air film thermal resistance, 0.62 °F·ft^2·h/Btu (0.109 m^2·K/W)

k_i = insulation thermal conductivity, °F·ft^2·h/Btu (m^2·K/W)

t_i = insulation thickness, in. (mm)

Equations 4-27, 4-28, and 4-29 can be used to determine the surface temperature for a given supply air temperature and insulation thickness or to determine the required insulation thickness for a given ambient dew-point temperature.

For ducts passing through conditioned space, including return air plenums, the insulation levels recommended in section 4.6.1 to limit heat gain will almost always be adequate to prevent condensation.

For ducts in unconditioned space, the required thickness should be determined based on the most extreme expected condition to be encountered. Equation 4-28 shows that the insulation requirement will be highest when the dew point is much higher than the supply air temperature and when the dew point is close to the dry-bulb temperature (high relative humidity).

Typical unconditioned spaces that would be exposed to conditions of high humidity include mechanical rooms ventilated by outside air or ceiling plenums that have ducted returns and that experience high infiltration rates. For such spaces, the extreme condensation condition can be estimated based on the design dew-point temperature (Colliver et al. 1995) at a dry-bulb temperature corresponding to 90% relative humidity.

Selecting insulation for a higher relative humidity will provide additional protection against extremely humid conditions. However, note that as the relative humidity approaches 100% and the dew point approaches the dry-bulb temperature, the required insulation thickness approaches infinity.

In most cases, temperatures in return ducts are greater than the ambient dew point surrounding the ducts, and insulation is not required. However, if the ambient dew point is expected to exceed the return air temperature, return duct insulation may be needed.

4.6.3 COVERAGE

The insulation must cover all duct and equipment surfaces that may be cooled below the ambient dew point. Fully insulate air-handling units from the mixed-air section to the supply air outlet. Supply ducts must be completely covered, especially at access doors. Requirements for insulation of outside air ductwork are the same as for conventional systems.

Seal penetrations in external insulation, such as for duct hangars, temperature control sensors, etc., with vapor retarder mastic or sealant. Cover weld pin fasteners with vapor retarder patches and a coating of mastic. For duct hangers with external insulation, provide a rigid insert of cellular glass or other noncompressive, nonconducting material to isolate the hangar from the duct and prevent overcompression of the insulation. One such arrangement is illustrated in Figure 4-4.

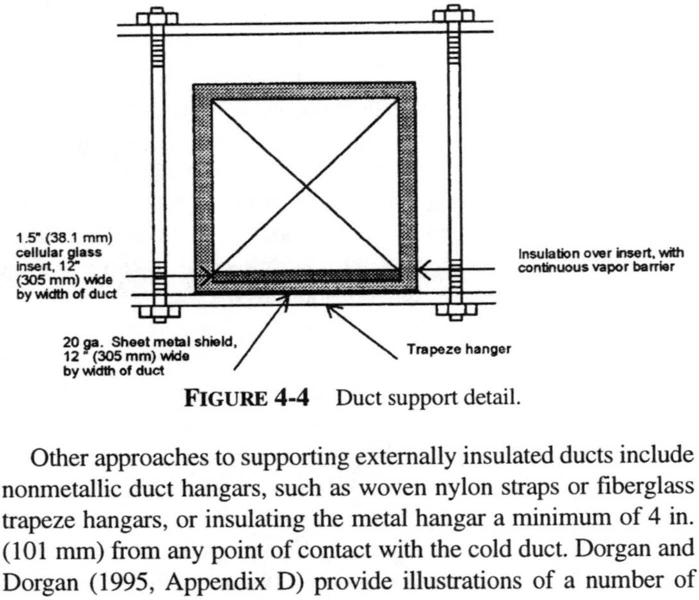

FIGURE 4-4 Duct support detail.

Other approaches to supporting externally insulated ducts include nonmetallic duct hangars, such as woven nylon straps or fiberglass trapeze hangars, or insulating the metal hangar a minimum of 4 in. (101 mm) from any point of contact with the cold duct. Dorgan and Dorgan (1995, Appendix D) provide illustrations of a number of duct insulation details for cold-air distribution systems.

When ducts are lined internally, ensure that the liner provides continuous, full coverage and that access doors are also lined. Installers of prefabricated double-wall insulated duct must take care that the sections are drawn up tightly together. If internal insulation is not continuous at the joints, ambient moisture will condense on the external surfaces.

Insulate all flexible duct connectors to the same thickness as adjacent duct insulation. Carry insulation on VAV box inlets to cover the entire inlet collar. Some VAV boxes are available with thermal isolation between the inlet collar and the box wall to prevent cooling of the external box surface by conduction.

Apply insulation after leakage tests have been completed but before the system is operated with supply air at the design supply temperature. Apply insulation only when the duct surfaces are dry and ambient temperature and humidity are within the manufacturer's recommended range for installation of the product.

4.6.4 VAPOR RETARDER

An effective vapor retarder is required to prevent moisture from penetrating insulation and condensing inside. This moisture will reduce the effectiveness of the insulation and can lead to corrosion of metal ducts. Wet insulation can also become a site for growth of microorganisms that can lead to degradation of indoor air quality.

When ducts are internally lined only, the duct walls together with the duct sealant must function as the vapor retarder. Seal all joints and seams to prevent moisture from entering the ducts when the system is not operating.

On external insulation, the vapor retarder must form a continuous seal with no breaks or punctures. Factory-applied vapor retarder jackets are typically specified to have a maximum permeance of 0.02 perm (1.15 ng/[s·m^2·Pa]) in accordance with ASTM Standard E96 (ASTM 1980). Cover all joints, seams, and penetrations in the factory-applied jacket with patches or strips of jacket material covered with vapor retarder coating. Field-applied jackets typically consist of two coats of vapor retarder coating, at least 1/16-in. (1.5 mm) thick, with a layer of glass cloth embedded between the coats.

In mechanical equipment rooms and other areas where the chances of damage to vapor retarders are increased, designers should specify closed-cell insulation, which is not susceptible to moisture damage.

Insulation specifications for cold-air distribution systems should emphasize the importance of providing a continuous, intact vapor retarder. Specifications should also stress that this item will receive particular attention in inspection and review for final acceptance. The general contractor and all other trades should be notified in the specifications and in construction meetings that damage to insulation will not be tolerated.

Training of operation and maintenance personnel should also include emphasis on the importance of maintaining the integrity of the vapor retarder.

4.6.5 DESIGN EXAMPLE—DETERMINE DUCT SURFACE TEMPERATURE

Determine the duct surface temperature for a duct with 1-in. external glass fiber insulation, carrying 45°F (7°C) air. The insulation thermal resistance is R = 3.0°F·ft^2·h/Btu (0.528 m^2·K/W). The ambient dry-bulb temperature is 78°F (26°C).

From Equation 4-27,

$$T_s = T_{sa} + \left[(T_a - T_{sa})\frac{R_i}{R_i + R_s}\right]$$

$$= 45 + \left[(78 - 45)\frac{3.0}{3.0 + 0.62}\right] \qquad \left(= 7 + \left[(26 - 7)\frac{0.528}{0.528 + 0.11}\right]\right)$$

$$= 72°F \qquad (= 22°C)$$

For the conditions given in this example, the 1-in. insulation considered here will prevent condensation whenever the ambient dew-point temperature is less than 72°F (22°C).

4.6.6 DESIGN EXAMPLE—DETERMINE REQUIRED INSULATION VALUE

Determine the required thermal resistance to prevent condensation on a duct carrying 45°F (7°C) air at ambient conditions of (a) 78°F (26°C) db, 50% RH, and (b) 78°F (26°C) db, 90% RH.

(a) The dew-point temperature at 78°F (26°C) db, 50% RH is 58°F (14°C). From Equation 4-28, the required thermal resistance is

$$R_i = R_s\frac{T_{dp} - T_{sa}}{T_a - T_{dp}}$$

$$= 0.62\frac{58 - 45}{78 - 58} \qquad \left[= 0.11\frac{14 - 7}{26 - 14}\right.$$

$$\left.= 0.07\,\text{m}^2 \cdot \text{K/W}\right]$$

$$= 0.40\ °\text{F} \cdot \text{ft}^2 \cdot \text{h/Btu}$$

For this example, with an ambient condition typical of a return air plenum, a minimal amount of insulation is needed to prevent condensation.

(b) The dew-point temperature at 78°F db (26°C), 90% RH, is 75°F (24°C). From Equation 4-28, the required thermal resistance is 6.2°F·ft^2·h/Btu (1.0 m^2·K/W).

From Table 4-6, this requirement would be satisfied by a 3-in. (76-mm), 0.75 lb/ft^3 (12 kg/m^3) or a 2-in. (51-mm), 1.5 lb/ft^3 (24 kg/m^3) external blanket. The requirement could also be met by combining an internal liner and an external blanket having the appropriate insulation values.

The ambient condition in this example represents an extreme case of condensation potential for many U.S. locations. The dew point of 75°F (24°C) is equal to or greater than the 1% design value for much of the U.S. (Colliver et al. 1995). The relative humidity of 90% represents an expected extreme condition in an unconditioned area exposed to outside air.

4.7 DUCT LEAKAGE

Leakage of supply air from ducts reduces the cooling capacity delivered to the conditioned space. If the leakage is to an unconditioned space, it represents an increase in the total load and a loss of energy from the system. If the leakage is to a conditioned space such as a return air plenum, the cooling capacity is not entirely lost from the system, but additional supply air must be provided to meet the space loads.

Leakage can also cause condensation on adjacent surfaces if the supply air temperature is lower than the ambient dew point. This concern is heightened with cold-air distribution systems.

The *HVAC Air Duct Leakage Test Manual* (SMACNA 1985) defines leakage classifications and provides a method for rating, analyzing, and testing duct leakage. The leakage class, C_L, is a number proportional to the leakage in cfm/100 ft^2 (L/s per m^2) of duct surface at 1 in. w.g. (250 Pa) static pressure.

ASHRAE/IESNA 90.1-1989 (ASHRAE 1989) recommends that ductwork designed to operate at static pressures above 3 in. w.g. (746 Pa) be constructed and leak-tested to leakage class 6 or better. A leakage class of 3 is attainable for all duct systems by careful selection of joints and sealing methods and by good workmanship (SMACNA 1990; ASHRAE 1993).

Although leakage as a percentage of total flow rate is a useful criterion for system analysis, the SMACNA leakage classifications and test method require evaluation in terms of leakage per unit of duct surface area. The leakage class can be related to a percentage of airflow if the system flow rate in cfm/ft^2 (L/s per m^2) of duct surface is known (ASHRAE 1993, chapter 32, Table 7).

Designers of air distribution systems should determine allowable leakage rates for each duct system and specify the appropriate pressure class, seal class, and acceptable leakage classification. Leakage from equipment must be accounted for separately from duct leakage. Specifications must define the amount and manner of required leakage testing and clearly indicate the acceptance criteria.

4.8 DUCT SIZING

Duct-sizing considerations for cold-air distribution systems are essentially the same as those for conventional designs. The reduction in supply air volume offers the designer more flexibility in selecting duct sizes. Typically, ducts are sized to the same pressure loss criteria used for conventional systems. This approach reduces duct size and cost, as well as fan energy, and often allows the use of round or oval ducts where space limitations would otherwise have required rectangular duct.

Round and oval ducts are inherently strong and rigid and are generally the most efficient and economical ducts for air distribution systems (ASHRAE 1996, chapter 16). Round and oval ducts are easier to install and to seal against leakage and have better sound control characteristics than rectangular ducts.

Cold-air duct systems may also be "oversized," using duct sizes typical of conventional systems. This oversizing reduces duct pressure loss, minimizes noise, and maximizes fan energy savings. However, the increased duct surface area per unit air volume leads to a higher supply air temperature rise than with standard sizing. This approach is particularly well suited to applications where noise control is a primary consideration, such as courthouses, theaters, and concert halls.

The trade-off between reducing duct size and reducing duct pressure loss is a design decision that depends on space limitations, ductwork costs, and energy costs. If the load requirements of a conventional system exceed the original design calculations, reducing the supply air temperature may be a means of increasing the distribution system capacity. This "safety factor" is not available for cold-air distribution systems, so designers should ensure that supply ducts are sized for the full anticipated future load.

A detailed discussion of duct-sizing methods is beyond the scope of this guide. Three common methods are equal friction, static regain, and the T-method. These methods are discussed in SMACNA (1990), ASHRAE (1993), Brooks (1995), and Tsal et al. (1988).

4.9 FAN SELECTION

Selection of supply fans for cold-air distribution systems is governed by the same considerations as for conventional designs. Guidance on fan selection is provided in the *1996 ASHRAE Handbook—HVAC Systems and Equipment* (ASHRAE 1996), in Lorsch et al. (1993), and in Carpenter (1984).

An important fan selection consideration is the placement of the fan relative to the cooling coil in either a "blow-through" or a "draw-through" configuration. In a blow-through configuration, the fan is upstream of the cooling coil. A draw-through configuration places the fan downstream of the coil.

With a draw-through arrangement, the fan heat is transferred to the supply airstream, increasing the supply air temperature by 2°F to 4°F (1°C to 2°C) and increasing the air volume required to meet the load. With a blow-through arrangement, the fan heat is transferred from the airstream directly to the coil, allowing a lower final supply air temperature from the same coil.

The blow-through fan configuration has some limitations that may offset the advantage of the small decrease in supply air temperature. A blow-through arrangement can produce nonuniform air distribution across the coil, which may be more likely to cause moisture carryover, and which may detract from coil performance. The less desirable fan inlet and exit conditions may produce a system effect that results in diminished fan performance.

In addition, a blow-through unit may be more likely to experience condensation on downstream surfaces, particularly if there is any fluctuation in the air temperature leaving the coil. For example, consider a blow-through system with bag filters downstream of the coil in which the coil discharge temperature varies between 48°F and 51°F (9°C and 11°C). With 48°F (9°C) discharge air, the metal filter frames are cooled to 48°F (9°C). If the discharge temperature rises to 51°F (11°C), moisture will condense on the 48°F (9°C) metal surfaces. Any of this moisture that finds its way into the absorbent filter medium becomes effectively trapped. It will not re-evaporate because the air passing over it is 100% saturated. Over time, enough moisture could build up to damage the filters, as well as providing a site for the growth of undesirable microorganisms. The fan heat added to the supply air in a draw-through arrangement provides a safety factor against the possibility of this downstream condensation.

If the lower supply temperature available from a blow-through configuration is an important design requirement, a flow straightener or a distance of three to five fan diameters between the fan and the coil is recommended.

While not directly related to the fan itself, designers selecting an air-handling unit configuration should ensure that outside air is thoroughly mixed with return air before entering the cooling coil. With low outside air temperatures, inadequate mixing can lead to freezing of cooling coils. This is an increased concern in cold-air distribution systems, in which outside air is a greater proportion of the total airflow. Designers of air-handling systems for cold climates should consider specifying air-blending devices or other means to ensure adequate mixing of outside and return air.

Blow-through fans improve the temperature uniformity of air reaching the cooling coil. However, severe temperature stratification can often persist even through a fan.

Normally, opposed-blade dampers are preferred to the parallel-blade arrangement for control of outside air and return air. Opposed-blade dampers have a more linear performance characteristic and help reduce temperature stratification. However, in some cases, mixing can be improved by arranging two sets of parallel blade dampers to blow toward each other.

Air-blending devices provide the best mixing performance but require a significant amount of space upstream and downstream.

4.10 PSYCHROMETRIC ANALYSIS

4.10.1 INTRODUCTION

The cold-air distribution system designer must perform a psychrometric analysis of the temperature and humidity conditions throughout the system. This analysis is essential to accurately determine system design requirements, particularly the air condition entering the cooling coil. The commonly used rules of thumb that often take the place of such an analysis are not applicable to cold-air distribution designs, where coil latent cooling capacity requirements must be carefully evaluated.

The calculations presented here are based on volume airflow rates expressed in terms of standard density. As discussed in section 4.1, actual airflow rates should be converted to standard flow rates for accurate calculations.

Psychrometric properties most pertinent to this analysis include dry-bulb and wet-bulb temperatures, relative humidity, and humidity ratio. Other properties are enthalpy, dew-point temperature, and specific volume. In general, if any two properties are known, the others can be calculated by any of several methods. Chapter 6 of the *1993 ASHRAE Handbook—Fundamentals* discusses methods for calculating psychrometric properties. A number of computer programs are on the market that perform psychrometric calculations. A psychrometric chart can be used to graphically determine properties, although the results are generally not as precise as with the numerical methods.

The analysis of heat transfer and psychrometric processes in an air distribution system generally proceeds as follows:

- Determine the air condition leaving the cooling coil.
- Determine the final supply air temperature to the space.
- Calculate the required supply airflow.
- Determine the return air condition entering the air-handling unit.
- Calculate the mixed air condition entering the coil.
- Evaluate the cooling coil loads.

A typical analysis requires the following input information:

- Space sensible load
- Space latent load
- Space temperature setpoint
- Supply air temperature leaving the cooling coil
- Fan temperature rise
- Supply duct temperature rise
- Supply duct leakage
- Outside air quantity
- Heat gain from lights to return air
- Heat gain from the roof to return air
- Infiltration into return plenum
- Ducted or plenum return and
- Draw-through or blow-through configuration.

Symbols and abbreviations used in this discussion are as follows:

db - dry-bulb temperature
wb - wet-bulb temperature
Δt - dry-bulb temperature difference
w - humidity ratio
RH - relative humidity
q - heat transfer
Q - airflow volume
oa - outside air
ma - mixed air
ra - return air
ea - exhaust air
ec - entering coil
lc - leaving coil
lf - leaving fan
sa - final supply air to space
rm - room or space
sens - room sensible
lat - room latent
inf - infiltration
leak - leakage
f( ) - psychometric property calculation

4.10.2 DETERMINE THE AIR CONDITION LEAVING THE COOLING COIL

The air condition leaving the cooling coil is determined by the required leaving coil dry-bulb temperature and the performance of the coil. For initial estimates before the coil is selected, the relative humidity leaving the coil can be assumed to be 95% to 99% RH. For standard designs with 55°F (13°C) leaving temperature, the relative humidity will be approximately 95%. At a temperature of 42°F (6°C) leaving the coil, the relative humidity will be 99%.

The humidity ratio is determined from the dry-bulb temperature and the relative humidity by a psychrometric property calculation.

$$w_{lc} = f(db_{lc}, rh_{lc}) \tag{4-30}$$

4.10.3 DETERMINE THE FINAL SUPPLY AIR TEMPERATURE TO THE SPACE

The final supply air temperature is determined from the coil leaving air temperature and an analysis of heat gains to the airstream before it is delivered to the space. The final supply temperature is equal to the leaving-coil temperature plus the distribution system temperature gains. If the fan is downstream of the coil in a draw-through configuration, the fan heat will cause a temperature rise. The temperature rise due to heat gain to the supply ducts, and from fan-powered mixing boxes if they are present, must also be added here.

The final supply air humidity ratio is equal to the humidity ratio leaving the coil, since there are no latent gains or losses in the distribution system.

$$db_{sa} = db_{lc} + \Delta T_{fan,draw\text{-}through} + \Delta T_{duct,\ sa} \quad (4\text{-}31)$$

$$W_{sa} = W_{lc} \quad (4\text{-}32)$$

4.10.4 CALCULATE THE REQUIRED SUPPLY AIRFLOW

The required supply air volume is calculated from the peak block space sensible load and the room-to-supply air temperature difference. The space loads are calculated as discussed under "Load Calculations" in section 4.3. The room temperature setpoint is selected as discussed in chapter 3.

$$Q_{sa} = \frac{q_{sens}}{C_1(db_{rm} - db_{sa})} \quad (4\text{-}33)$$

where

$C_1 = 1.1$ Btu·min/ft³·h·°F (1.24 W·s/m³·°C)

The space humidity condition is calculated from the supply air condition and the space latent load.

$$W_{rm} = W_{lc} + \frac{q_{lat}}{C_2 Q_{sa}} \quad (4\text{-}34)$$

where

$C_2 = 4840$ Btu·min·lb_a/h·ft³·lb_w (3002 W·s/L)

$$rh_{rm} = f(db_{rm}, W_{rm}) \quad (4\text{-}35)$$

4.10.5 DETERMINE THE RETURN AIR CONDITION ENTERING THE AIR-HANDLING UNIT

The final return air condition is determined from the room condition and from heat gains and losses in the return air plenum. For ducted returns, heat transfer to the return air may be calculated by the methods given in section 4.5. However, the temperature difference between the return duct and the surrounding space is usually so small that there is negligible heat transfer through the duct wall.

If exhaust air to balance the outside air intake is removed directly from the space, the return air volume is less than the supply volume.

$$Q_{ra} = Q_{sa} - Q_{ea,space} \tag{4-36}$$

Heat gains to return air in a plenum include that portion of the heat of lights that is transferred directly to the return air and the roof cooling load if the return air plenum is located below a roof.

$$\Delta T_{lights} = \frac{q_{lights}}{C_1 Q_{ra}} \tag{4-37}$$

$$\Delta T_{roof} = \frac{q_{roof}}{C_1 Q_{ra}} \tag{4-38}$$

The return air also loses heat to the supply air ducts, equivalent to the heat gained by the supply ducts from the plenum. When the return air volume is less than the supply volume, the temperature loss in the return air is proportionately greater than the supply air rise.

$$\Delta T_{duct,\,ra} = -\Delta T_{duct,\,sa}\frac{Q_{sa}}{Q_{ra}} \tag{4-39}$$

The return air temperature is affected by leakage from supply ducts and by infiltration to the return air plenum.

$$\Delta T_{leak} = \frac{q_{leak}}{Q_{ra}}(db_m - db_{sa}) \tag{4-40}$$

$$\Delta T_{ra,\,infil} = \frac{q_{infil}}{Q_{ra}}(db_{oa} - db_{rm}) \tag{4-41}$$

If the system includes a return fan, the return air will experience a temperature rise from the fan heat, which can be calculated using the equations in section 4.4.2.

The final return air temperature is calculated from the room temperature and the sum of the return air gains.

$$db_{ra} = db_{rm} + \Delta T_{lights} + \Delta T_{roof} + \Delta T_{duct,ra} + \Delta T_{leak} + \Delta T_{infil} + \Delta T_{retfan} \tag{4-42}$$

The return air humidity ratio is equal to the room humidity ratio plus any latent gains or losses from supply duct leakage and infiltration of outside air into the return plenum.

$$W_{ra} = W_{rm} + \frac{Q_{leak}}{Q_{ra}}(W_{sa} - W_{rm}) + \frac{Q_{inf}}{Q_{ra}}(W_{oa} - W_{rm}) \quad (4\text{-}43)$$

4.10.6 CALCULATE THE MIXED-AIR CONDITION ENTERING THE COIL

The mixed-air condition is calculated from the return air condition, the outside air condition, and the volumes of return and outside air. The return air volume is increased by the amount of supply duct leakage to the return plenum. If exhaust air to balance the outside air intake is removed at the air-handling unit, the final return air volume is reduced.

Note that the outside airflow in particular should be expressed in terms of standard airflow because the outside air density will differ most from standard and will have the greatest effect on the accuracy of the calculation.

$$Q_{ra,final} = Q_{ra} + Q_{leak} - Q_{ea,ahu} \quad (4\text{-}44)$$

$$db_{ma} = \frac{Q_{oa}db_{oa} + Q_{ra,final}db_{ra}}{Q_{ra,final} + Q_{oa}} \quad (4\text{-}45)$$

$$W_{ma} = \frac{Q_{oa}W_{oa} + Q_{ra,final}W_{ra}}{Q_{ra,final} + Q_{oa}} \quad (4\text{-}46)$$

If the supply fan is upstream of the coil in a blow-through configuration, the temperature entering the coil is increased by the fan temperature rise. The humidity ratio entering the coil is equal to the mixed-air humidity ratio. The wet-bulb temperature entering the coil is calculated from the dry-bulb and humidity ratio.

$$db_{ec} = db_{ma} + \Delta T_{fan,\ blow\text{-}through} \quad (4\text{-}47)$$

$$W_{ec} = W_{ma} \quad (4\text{-}48)$$

$$wb_{ec} = f(db_{ec}, W_{ec}) \quad (4\text{-}49)$$

4.10.7 EVALUATE THE COOLING COIL LOADS

With the entering air condition established, the designer can select a cooling coil to provide the required performance. The coil manufacturer provides data on the leaving air humidity condition for the selected coil at the given conditions. If the leaving air humidity is significantly different from the initial assumption, it will be necessary to iterate the calculation procedure.

The sensible, latent, and total cooling coil loads are calculated as follows. Note that the total air volume through the coil is equal to the supply air quantity delivered to the space plus the supply duct leakage.

$$q_{sens} = C_1(Q_{sa} + Q_{leak})(db_{ec} - db_{lc}) \tag{4-50}$$

$$q_{\mathrm{lat}} = C_2(Q_{\mathrm{sa}} + Q_{\mathrm{leak}})(W_{\mathrm{ec}} - W_{\mathrm{lc}}) \tag{4-51}$$

$$\begin{aligned} q_{total} &= q_{sens} + q_{lat} \\ &= C_{13}(Q_{sa} + Q_{leak})(h_{ec} - h_{lc}) \end{aligned} \tag{4-52}$$

where

C_1 = 1.1 Btu·min/ft^3·h·°F (1.24 W·s/m^3·°C)

C_2 = 4840 Btu·min·lb$_a$/h·ft^3·lb$_w$ (3002 W·s/L)

C_{13} = 4.5 lb·min/ft^3·h (1.2 kg/m^3)

4.10.8 DESIGN EXAMPLE—PSYCHROMETRIC ANALYSIS, 55°F (13°C) SUPPLY AIR

Calculate the air condition entering the cooling coil for a system in a draw-through configuration with a plenum return, with the following input parameters:

Space sensible load:	350,000	Btu/h (102,585 W)
Space latent load:	55,995	Btu/h (16,399 W)
Space temperature setpoint:	75°F	(23.9°C)
Temperature leaving cooling coil:	55°F	(12.8°C)
Fan temperature rise:	2.1°F	(1.2°C)
Supply duct temperature rise:	2.3°F	(1.3°C)
Supply duct leakage	1,000	cfm (472 L/s)
Outside air quantity:	3,000	cfm (1,416 L/s)
Return air heat gain from lights (38,700-W total, 20% to plenum):	26,420	Btu/h (7,740 W)
Return air heat gain from roof:	32,000	Btu/h (9,380 W)
Infiltration to return plenum:	100	cfm (47.2 L/s)
Leaving coil relative humidity:	95%	(assumed)

Humidity ratio of outside air:

$$W_{oa} = f(91°\text{F } db, 75°\text{F} wb) = 0.0141 \text{lb}_\text{w}/\text{lb}_\text{a} \qquad \begin{pmatrix} W_{oa} = f(32.8°\text{C } db, 23.3°\text{C } wb) \\ = 0.0141 \text{kg}_\text{w}/\text{kg}_\text{a} \end{pmatrix}$$

Humidity ratio leaving the coil:

$$W_{lc} = f(55°\text{F } db, 95\%\ rh) = 0.0087\ \text{lb}_\text{w}/\text{lb}_\text{a} \qquad \begin{pmatrix} W_{lc} = f(12.8°\text{C } db, 95\%\ rh) \\ = 0.0087(0.0141)\text{kg}_\text{w}/\text{kg}_\text{a} \end{pmatrix}$$

Final supply air condition:

$$db_{sa} = 55 + 2.1 + 2.3 = 59.4°\text{F}$$

$$W_{lc} = W_{sa} = 0.0087\ \text{lb}_\text{w}/\text{lb}_\text{a}$$

$$\begin{pmatrix} db_{sa} = 12.8 + 1.2 + 1.3 \\ = 15.3°\text{C} \\ W_{sa} = W_{lc} = 0.0087\text{kg}_\text{w}/\text{kg}_\text{a} \end{pmatrix}$$

Required supply air volume:

$$Q_{sa} = \frac{350,000}{1.1(75 - 59.4)} = (20,396\ \text{cfm}) \qquad \begin{pmatrix} Q_{sa} = \frac{102,585}{1.24(23.9 - 15.3)} \\ = 9,620\ \text{L/s} \end{pmatrix}$$

Space humidity:

$$W_{rm} = 0.0087 + \frac{55,995}{(4840)(20,396)} = 0.0093\ \text{lb}_\text{w}/\text{lb}_\text{a}$$

$$\text{RH}_{rm} = f(75°\text{F } db, 0.0093\ \text{lb}_\text{w}/\text{lb}_\text{a}) = 50\%$$

$$\begin{pmatrix} W_{rm} = 0.0087 + \frac{16,399}{(3002)(9,620)} = 0.0093\ \text{kg}_\text{w}/\text{kg}_\text{a} \\ \text{RH}_{rm} = f(23.9°\text{C } db, 0.0093\ \text{kg}_\text{w}/\text{kg}_\text{a}) = 50 \end{pmatrix}$$

Final return air condition:

$$Q_{ra} = 20,396 - 3,000 = 17,396\ \text{cfm} \qquad \begin{pmatrix} Q_{ra} = 9,620 - 1,416 \\ = 8,204\ \text{L/s} \end{pmatrix}$$

$$\Delta T_{leak} = \frac{-1000}{1,396}(75 - 59.4) = -0.9°F$$

$$\left(\Delta T_{leak} = \frac{-472}{8,204}(23.9 - 15.3) = -0.5°C\right)$$

$$\Delta T_{lights} = \frac{26,420}{(1.1)(17,396)} = 14°F$$

$$\left(\Delta T_{lights} = \frac{7,740}{(1.2)(8,204)} = 0.8°C\right)$$

$$\Delta T_{ra,\,infil} = \frac{100}{17,396}(91 - 75) = 0.1°F$$

$$\left(\Delta T_{ra,\,infil} = \frac{47.2}{8,226}(32.8 - 23.9) = 0.1°C\right)$$

$$\Delta T_{roof} = \frac{32,000}{(1.1)(17,396)} = 1.7°F$$

$$\left(\Delta T_{roof} = \frac{9,379}{(1.2)(8,204)} = 0.9°C\right)$$

$$\Delta T_{duct,\,ra} = 1.8\frac{20,396}{17,396} = -2.7°F$$

$$\left(\Delta T_{duct,\,ra} = 1.0\frac{9,620}{8,204} = -1.5°C\right)$$

$$db_{ra} = 75 + 1.4 + 0.1 + 1.7 - 0.9 - 2.7 = 74.6°F$$

$$W_{ra} = 0.0093 + \frac{1000}{17,396}(0.0087 - 0.0093) + \frac{100}{17,396}(0.0141 - 0.0093) = 0.0093\ lb_w/lb_a$$

$$\left(db_{ra} = 23.9 + 0.8 + 0.1 + 0.9 - 0.5 - 1.5 = 23.7°C\right.$$

$$\left.W_{ra} = 0.0093 + \frac{472}{8,204}(0.0087 - 0.0093) + \frac{47.2}{8,204}(0.0141 - 0.0093) = 0.0093\ kg_w/kg_a\right)$$

Mixed air condition:

$$db_{ma} = \frac{3,000(91.0) + 18,396(74.6)}{3,000 + 18,396} = 76.9°F$$

$$\left(db_{ma} = \frac{1,416(32.8) + 8,676(23.7)}{8,676 + 1,416} = 25.0°C\right)$$

$$W_{ma} = \frac{3,000(0.0141) + 18,396(0.0093)}{18,396 + 3,000} = 0.0100\ lb_w/lb_a$$

$$\left(W_{ma} = \frac{1,416(0.0141) + 8,676(0.0093)}{8,676 + 1,416} = 0.0100\ kg_w/kg_a\right)$$

Entering coil condition, draw-through:

$$\begin{aligned} db_{ec} &= db_{ma} = 76.9°\text{F} \\ W_{ec} &= W_{ma} = 0.0100\ \text{lb}_\text{w}/\text{lb}_\text{a} \\ wb_{ec} &= f(W_{ec}, db_{ec}) \\ &= f(0.0100\ \text{lb}_\text{w}/\text{lb}_\text{a}\ 76.9°\text{F}) \\ &= 64.3°\text{F} \end{aligned}$$

$$\left(\begin{aligned} db_{ec} &= db_{ma} = 25.0°\text{C} \\ W_{ec} &= W_{ma} = 0.0100\ \text{kg}_\text{w}/\text{kg}_\text{a} \\ wb_{ec} &= f(W_{ec}, db_{ec}) \\ &= f(0.0100\ \text{kg}_\text{w}/\text{kg}_\text{a}\ 25.0°\text{C}) \\ &= 18.0°\text{C} \end{aligned}\right)$$

Cooling coil loads:

$$q_{sens} = 1.1(20{,}396 + 1000)(76.9 - 55) = 515{,}430\ \text{Btu/h}$$
$$(q_{sens} = 1.24(9{,}620 + 472)(25.0 - 12.8) = 152{,}671\ \text{W})$$

$$q_{lat} = 4840(20{,}396 + 1{,}000)(0.0100 - 0.0087) = 134{,}624\ \text{Btu/h}$$
$$(q_{lat} = 3002(9{,}620 + 472)(0.0100 - 0.0087) = 39{,}385\ \text{W})$$

$$q_{total} = 515{,}430 + 134{,}624 = 650{,}054\ \text{Btu/h}$$
$$(q_{total} = 152{,}671 + 39{,}385 = 192{,}056\ \text{W})$$

The psychrometric process for this example is plotted in Figure 4-5.

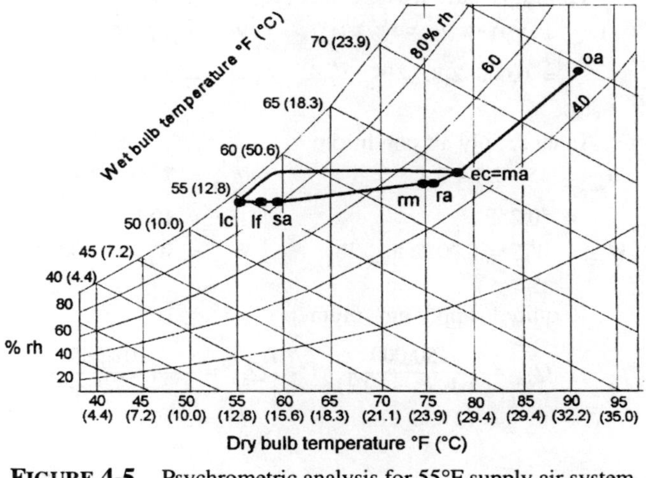

FIGURE 4-5 Psychrometric analysis for 55°F supply air system.

4.10.9 DESIGN EXAMPLE—PSYCHROMETRIC ANALYSIS, 45°F (7°C) SUPPLY AIR

Calculate the air condition entering the cooling coil for a system in a draw-through configuration with a plenum return with the following input parameters:

Space sensible load:	350,000	Btu/h (102,585 W)
Space latent load:	69,569	Btu/h (20,372 W)
Space temperature setpoint:	75	°F (23.9°C)
Temperature leaving cooling coil:	45	°F (7.2°C)
Fan temperature rise:	2.1	°F (1.2°C)
Supply duct temperature rise:	3.1	°F (1.7°C)
Supply duct leakage	600	cfm (283 L/s)
Outside air quantity:	3,000	cfm (1,416 L/s)
Return air heat gain from lights (38,700 W total, 20% to plenum):	26,420	Btu/h (7,740 W)
Return air heat gain from roof:	32,000	Btu/h (9,379 W)
Infiltration to return plenum:	100	cfm (47.2 L/s)
Leaving coil relative humidity:	99%	(assumed)

Humidity ratio of outside air:

$$W_{lc} = f(45°\text{F } db, 74°\text{C } wb) = 0.0141 \text{ lb}_\text{w}/\text{lb}_\text{a} \qquad \left(W_{oa} = f(32.8°\text{C } db, 23.3°\text{C } wb) = 0.0141 \text{ kg}_\text{w}/\text{kg}_\text{a}\right)$$

Humidity ratio leaving the coil:

$$W_{oa} = f(91°\text{F } db, 99\% \text{ RH}) = 0.00062 \text{ lb}_\text{w}/\text{lb}_\text{a} \qquad \left(W_{lc} = f(72°\text{C } db, 99\% \text{ RH}) = 0.0062 \text{ kg}_\text{w}/\text{kg}_\text{a}\right)$$

Final supply air condition:

$$db_{sa} = 45 + 2.1 + 3.1 = 50.2°\text{F} \qquad \left(db_{sa} = 7.2 + 1.2 + 1.7 = 10.1°\text{C}\right)$$

$$W_{sa} = W_{lc} = 0.0062 \text{ lb}_\text{w}/\text{lb}_\text{a} \qquad \left(W_{sa} = W_{lc} = 0.0062 \text{ kg}_\text{w}/\text{kg}_\text{a}\right)$$

Required supply air volume:

$$Q_{sa} = \frac{350{,}000}{1.1(75 - 50.2)} = 12{,}830 \text{ cfm} \qquad \left(Q_{sa} = \frac{102{,}585}{1.24(23.9 - 10.1)} = 5{,}995 \text{ L/s}\right)$$

Space humidity:

$$W_{rm} = 0.0062 + \frac{69{,}569}{(4840)(12{,}830)} = 0.0073\ \text{lb}_w/\text{lb}_a$$

$$RH_{rm} = f(75°\text{F } db, 0.0073\ \text{lb}_w/\text{lb}_a) = 40\%$$

$$\left(\begin{aligned} W_{rm} &= 0.0062 + \frac{20{,}372}{(3002)(5{,}995)} = 0.0073\ \text{kg}_w/\text{kg}_a \\ RH_{rm} &= f(23.9°\text{C } db, 0.0073\ \text{kg}_w/\text{kg}_a) = 40\% \end{aligned}\right)$$

Final return air condition:

$$\begin{aligned} Q_{ra} &= 12{,}830 - 3{,}000 \\ &= 9{,}830\ \text{cfm} \end{aligned} \qquad \left(\begin{aligned} Q_{ra} &= 5{,}995 - 1{,}416 \\ &= 4{,}579\ \text{L/s} \end{aligned}\right)$$

$$\begin{aligned} \Delta T_{leak} &= \frac{-600}{9,830}(75 - 50.2) \\ &= -1.5°\text{F} \\ \Delta T_{lights} &= \frac{26{,}420}{(1.1)(9,830)} \\ &= 2.4°\text{F} \\ \Delta T_{ra,\ infil} &= \frac{100}{9{,}830}(91 - 75) \\ &= 0.2°\text{F} \\ \Delta T_{roof} &= \frac{32{,}000}{(1.1)(9{,}830)} \\ &= 3.0°\text{F} \\ \Delta T_{duct,\ ra} &= -3.1\frac{12{,}830}{9{,}830} \\ &= -4.0°\text{F} \end{aligned} \qquad \left(\begin{aligned} \Delta T_{leak} &= \frac{-283}{6,050}(23.9 - 10.1) \\ &= -0.9°\text{C} \\ \Delta T_{lights} &= \frac{7,740}{(1.2)(4,579)} \\ &= 1.4°\text{C} \\ \Delta T_{ra,\ infil} &= \frac{47.2}{5,995}(32.8 - 23.9) \\ &= 0.1°\text{C} \\ \Delta T_{roof} &= \frac{9{,}379}{(1.2)(4,579)} \\ &= 1.7°\text{C} \\ \Delta T_{duct,\ ra} &= -1.7\frac{5{,}995}{4,579} \\ &= -2.2°\text{C} \end{aligned}\right)$$

$$db_{ra} = 75 + 2.4 + 3.0 + 0.2 - 1.5 - 4.0 = 75.1°\text{F}$$

$$W_{ra} = 0.0073 + \frac{1000}{9{,}830}(0.0062 - 0.0073) + \frac{100}{9{,}830}(0.0141 - 0.0073) = 0.0073\ \text{lb}_w/\text{lb}_a$$

$$\left(\begin{aligned} db_{ra} &= 23.9 + 1.4 + 1.7 + 0.1 - 0.9 - 2.2 = 24.0°\text{C} \\ W_{ra} &= 0.0073 + \frac{472}{4{,}579}(0.0087 - 0.0073) \\ &\quad + \frac{47.2}{4{,}579}(0.0141 - 0.0073) = 0.0073\ \text{kg}_w/\text{kg}_a \end{aligned}\right)$$

Mixed-air condition:

$$db_{ma} = \frac{3{,}000\ (91.0) + 10{,}430\ (75.0)}{3{,}000\ + 10{,}430} = 78.7°\text{F}$$

$$\left(db_{ma} = \frac{1{,}416(32.8) + 4{,}862(24.0)}{4{,}862 + 1{,}416} = 26.0°\text{C}\right)$$

$$W_{ma} = \frac{3{,}000\ (0.0141) + 10{,}430\ (0.0073)}{10{,}430\ + 3{,}000} = 0.0088\ \text{lb}_\text{w}/\text{lb}_\text{a}$$

$$\left(W_{ma} = \frac{1{,}416(0.0141) + 4{,}862(0.0073)}{4{,}862\ + 1{,}416} = 0.0088\ \text{kg}_\text{w}/\text{kg}_\text{a}\right)$$

Entering coil condition, draw-through:

$$\begin{aligned} db_{ec} &= db_{ma} = 78.7°\text{F} \\ W_{ec} &= W_{ma} = 0.0088\ \text{lb}_\text{w}/\text{lb}_\text{a} \\ wb_{ec} &= f(W_{ec}, db_{ec}) \\ &= f(0.0088\ \text{lb}_\text{w}/\text{lb}_\text{a}, 78.7°\text{F}) \\ &= 63.2°\text{F} \end{aligned}$$

$$\left(\begin{aligned} db_{ec} &= db_{ma} = 26.0°\text{C} \\ W_{ec} &= W_{ma} = 0.0088\ \text{kg}_\text{w}/\text{kg}_\text{a} \\ wb_{ec} &= f(W_{ec}, db_{ec}) \\ &= f(0.0088\ \text{kg}_\text{w}/\text{kg}_\text{a}, 26.0°\text{C}) \\ &= 17.3°\text{C} \end{aligned}\right)$$

Cooling coil loads:

$$q_{sens} = 1.1(12{,}830\ + 600)(78.7 - 45) = 497{,}850\ \ \text{Btu/h}$$
$$(q_{sens} = 1.24(5{,}995\ + 283)(26.0 - 7.2) = 146{,}357\ \ \text{W})$$

$$q_{lat} = 4840(12{,}830\ + 600)(0.0088 - 0.0062) = 169{,}003\ \ \text{Btu/h}$$
$$(q_{lat} = 3002(5{,}995\ + 283)(0.0088 - 0.0062) = 49{,}002\ \ \text{W})$$

$$q_{total} = 497{,}850\ + 169{,}003\ = 666{,}853\ \ \text{Btu/h}$$
$$(q_{total} = 146{,}357\ + 49{,}002\ = 195{,}359\ \ \text{W})$$

The psychrometric process for this example is plotted in Figure 4-6.

4.11 COOLING COIL SELECTION

4.11.1 GENERAL CONSIDERATIONS

Cooling coil selection takes on increased importance in cold-air distribution systems because many design parameters differ from typical designs. A cold-air coil selection will have some or all of the following differences compared to conventional selections:

- lower leaving air temperature,
- lower entering fluid temperature,

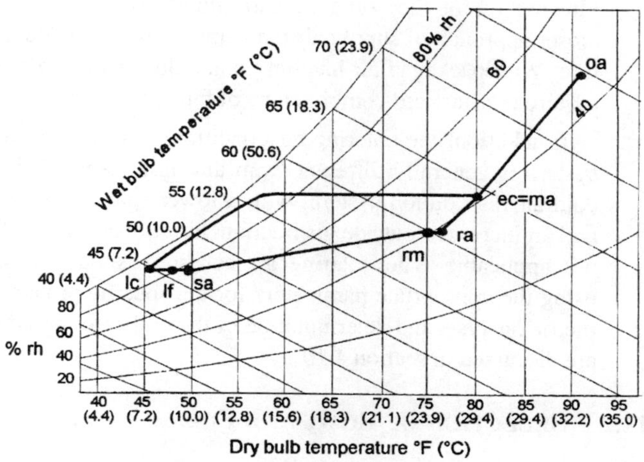

FIGURE 4-6 Psychrometric analysis for 45°F supply air system.

- lower face velocity,
- closer approach between the entering chilled-fluid temperature and the leaving air temperature, and
- wider range between the chilled-fluid supply and return temperatures.

Common design parameters for conventional and cold-air distribution systems are given in Table 4-9.

TABLE 4-9 COMMON COIL SELECTION PARAMETERS

	Standard	Cold Air Distribution
Leaving air temperature, °F (°C)	55 (12.8)	42-50 (5.6-10.0)
Chilled-fluid temperature entering coil, °F (°C)	42-45 (5.6-7.2)	36-42 (2.2-5.6)
Face velocity, fpm (m/s)	450-550 (2.3-2.8)	300-450 (1.5-2.3)
Approach, °F (°C)	10-13 (5.5-7.2)	4-10 (2.2-5.5)
Chilled-fluid temperature range, °F (°C)	10-16 (5.5-8.8)	16-24 (8.8-13.2)

Coils for cold-air distribution systems are generally selected with lower face velocities, 300 to 450 fpm (1.5 to 2.3 m/s), than the 450 to 550 fpm (2.3 to 2.8 m/s) typically used in conventional designs. The lower face velocity reduces the chance of condensate blowing off of the coil face and allows coils to be selected for closer approach of supply air temperature to chilled-fluid temperature. With packaged air-handling units, this design parameter may require nonstandard combinations of fan and cabinet sizes.

In addition, the entering air condition in a cold-air distribution system is generally different from that in a standard design. In a cold-air distribution system, with a lower space relative humidity and an increased outside air fraction, traditional rules of thumb are not applicable. The entering air condition should be calculated using the appropriate parameters for the specific project. Psychrometric analysis and determination of the coil entering air condition are discussed in section 4.10.

4.11.2 SELECTION OF ROWS AND FINS

Coil selections for conventional designs typically have 4 or 6 rows, with 8 to 14 fins per inch (0.32 to 0.55 fins per mm). Cooling coils for cold-air distribution systems require more heat transfer surface area because more heat is being transferred from each unit of air passing through the coil. Coil selections for cold-air systems may have from 6 to 12 rows, with up to 14 fins per inch (0.55 fins per mm).

Many designers prefer to limit their selections to no more than 8 rows to make coil cleaning easier and to minimize air and water pressure drops. Fin spacings of no more than 8 to 10 fins per inch (0.39 to 0.47 fins per mm) minimize the chances of condensed moisture blowing off of coils into the airstream.

If extreme approaches or ranges are not required, it is not difficult to select coils within these limits. However, coil selection becomes more difficult if an approach of less than 6°F (3.3°C) or a fluid temperature range of more than 24°F (13.2°C) is required. If both a close approach and a wide temperature range are desired, the number of possible coil selections is even more limited, and it may be necessary to select 10- or 12-row coils.

Multiple or split coils are often used to facilitate cleaning in applications requiring coils with 12 or more rows. For example, a 12-row coil is replaced by two 6-row coils. The two coils are spaced to allow access for cleaning or installed to permit individual removal.

Special coatings for fins are available to reduce the tendency for moisture carryover. This option may be applicable for extremely demanding selections where closely spaced fins increase the possibility of carryover. Before specifying a coated surface, designers should carefully evaluate the expected life of the coating and its impact on coil cleaning.

Textured internal surfaces on coil tubes can increase fluid turbulence and enhance fluid-side heat transfer, allowing coils to provide the required performance with fewer rows and fins. This reduces coil cost and air pressure drop but increases fluid pressure drop. A similar effect can sometimes be achieved by selecting coils with smaller diameter tubes.

4.11.3 COIL CIRCUITING

The heat transfer performance of a coil is partly determined by the arrangement of fluid circuits in the coil.

Coils are composed of horizontal tubes running across the width of the coil, arranged in vertical *rows*. The number of continuous flow paths from the fluid supply to return, or *circuits,* is determined by the arrangement of tube connections at the ends of the rows. Each portion of the fluid flow makes a number of *passes* across the width of the coil. The number of circuits in a coil is also referred to as the *serpentine* arrangement.

These coil circuiting parameters are related by the following expression:

Number of rows = Number of circuits × Number of passes

With standard or full-coil circuiting, each row of tubes carries the full flow of water in one pass across the width of the coil.

A double-circuit, or double-serpentine, coil contains twice as many tube circuits as a full-circuit coil. Each row carries half of the fluid flow, and any given volume of the fluid flows through only half of the rows.

Partial-circuit, or partial-serpentine, coils contain fewer circuits, with multiple passes in each row. For example, in a half-circuit coil, half the tubes in each row carry the full flow across the coil and the other half carry it back in the other direction.

Standard-, double-, and half-circuiting arrangements are illustrated in Figure 4-7.

Standard-circuit coils are selected for most cold-air distribution applications. Multiple-circuit coils are designed for high fluid flow and low temperature range and are generally not recommended for cold-air distribution. While some multiple-circuit coil selections provide acceptable performance at full-load conditions, tube velocities are unacceptably low when fluid flow is reduced at partial loads.

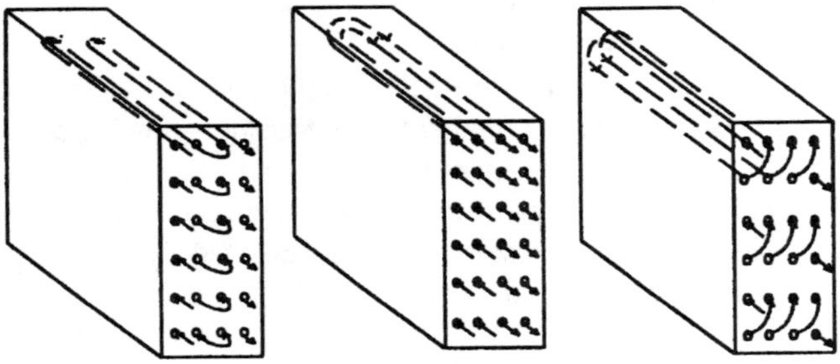

FIGURE 4-7 Coil circuit arrangements.

Partial-circuiting arrangements can be used in cold-air distribution systems to provide closer approaches and wider temperature ranges. In partial-circuit coils, heat transfer performance is improved by the longer flow path through the coil and by the increased fluid velocity and turbulence. However, with this improved heat transfer comes an increase of 100% or more in fluid pressure drop. Partial-circuit coils are particularly applicable when low tube velocity at partial loads is a concern.

4.11.4 COIL PERFORMANCE RATINGS

Many cooling coil manufacturers provide performance ratings certified under ARI Standard 410-91 (ARI 1991). This standard provides a method for extending laboratory test data to operating conditions and coil configurations other than those tested.

ARI Standard 410-91 covers performance ratings to a minimum fluid velocity of 1 fps (0.3 m/s) and a minimum fluid-side Reynolds number of 3,100. Coils for cold-air distribution systems are often selected with tube velocities at or below this level. There is some evidence that the algorithms used under Standard 410-91 overestimate coil performance at low fluid velocities. Mirth et al. (1993) found that the manufacturer's software overpredicted the heat transfer rate of tested coils by as much as 8% at a Reynolds number of 3,100. The overprediction ranged from 12% to 18% at a Reynolds number of 2,300.

Designers applying coils near or below tube velocities of 1 fps (0.3 m/s) may wish to modify their selections or allow a safety factor in the specification of coil performance. Tube velocities can be increased by using smaller diameter tubes or partial-circuit coils.

Internal tube enhancement improves heat transfer at low tube velocities. One method of providing a safety factor is to specify coils capable of providing the required capacity at an entering fluid temperature 2°F (1°C) higher than the design supply temperature.

4.11.5 PRESSURE DROPS

A given heat transfer requirement can generally be met with any of a number of coil selections. There is normally a trade-off between air- and fluid-side pressure drops. For example, heat transfer performance can be improved by providing more closely spaced fins, which increases air pressure drop. The same performance could be achieved with fewer fins but with additional rows. This generally reduces the air pressure drop but increases fluid pressure drop. Internal tube enhancement improves heat transfer with no effect on air pressure drop but a significant increase in fluid pressure drop.

Various fin configurations are also available to emphasize heat transfer performance at the cost of increased air pressure drop or to minimize pressure drop at the cost of additional heat transfer surface.

The selection of number of rows and fins and the trade-offs among air and fluid pressure drop are generally resolved by an economic analysis of fan and pump operating costs and equipment first cost. Many designers set design goals for maximum pressure drops for conventional selections, which may need to be increased for cold-air distribution systems.

Table 4-10 lists typical pressure drops for cooling coils selected to provide leaving air temperatures of 55°F (12.8°C), 45°F (7.2°C), and 42°F (5.6°C).

When coils for cold-air distribution applications are selected for face velocities of 300 fpm (1.5 m/s) or lower, the pressure drops are lower than the typical values cited in Table 4-10 and may be equivalent to or less than those for conventional designs.

TABLE 4-10 TYPICAL COIL PRESSURE DROPS

Supply Air Temperature		Air Pressure Drop		Fluid Pressure Drop	
°F	°C	in. w.g.	Pa	ft	m
55	12.8	0.5 - 1.0	125 - 250	6 - 20	1.8 - 6
45	7.2	0.6 - 1.0	150 - 250	9 - 25	2.7 - 7.6
42	5.6	0.6 - 1.3	150 - 320	13 - 30	4 - 9

Even when coil pressure drops are increased, these increases typically do not result in significant energy penalties. Because of the larger temperature differences present in the air and water distribution systems, fan and pump energy are still substantially lower than in conventional designs. Further, a high fluid-side pressure drop at design flow improves flow control over the range of control valve modulation.

4.11.6 GLYCOL CONSIDERATIONS

Glycol heat transfer fluids have higher viscosity, lower heat capacity, and lower thermal conductivity than water. More glycol solution must be circulated through a given coil to achieve the same heat transfer rates as with water. The increased flow rate and viscosity lead to higher fluid pressure drops. Additional heat transfer surface is needed, leading to higher air pressure drops.

For comparable coils with the same entering fluid temperature, the air pressure drop with ethylene glycol can be expected to be 0.1 to 0.2 in. w.g. (25 to 50 Pa) higher than that with water. The fluid pressure drop will be 3 to 6 ft (0.9 to 1.8 m) higher. Propylene glycol fluids are more viscous, and the fluid pressure drop will be 4 to 12 ft (1.2 to 3.7 m) higher than for water. Reducing the entering fluid temperature by 1°F to 2°F (0.55°C to 1.1°C) can eliminate the air-side penalty, but there may still be an increased fluid-side pressure drop of 2 to 6 ft (0.6 to 1.8 m), depending on the selection and on glycol concentration and type.

4.11.7 PART-LOAD PERFORMANCE

Part-load performance should also be considered in selecting a coil. Many cold-air distribution systems, especially those with close approaches or high temperature rises, operate with relatively low fluid-side tube velocities compared to conventional selections. At partial loads, when tube velocities are reduced even further, the fluid flow may become laminar, sharply reducing heat transfer performance. When this occurs, the discharge temperature increases and the cooling coil valve opens. The fluid flow again becomes turbulent, the heat transfer rate increases, and the discharge temperature drops. Discharge air temperature control may become unstable as the coil valve opens and closes in an attempt to deliver the appropriate fluid flow rate.

This instability at low loads can be avoided by using internal tube enhancement to increase fluid-side heat transfer, by using smaller-diameter tubes, or by selecting partial-circuited coils that operate with higher tube velocities. These options increase the coil's fluid-side pressure drop. However, this increase may be acceptable where an increased fluid temperature range provides substantial reductions in pumping power.

4.11.8 CONDENSATE HANDLING

Cooling coil selection and configuration for cold-air distribution should account for the increased flow of condensate. With stacked coil arrangements, provide intermediate condensate pans, as illustrated in Figure 4-8. The intermediate drain pan prevents moisture carryover that might otherwise occur if the condensate from upper coils flows down the face of lower coils.

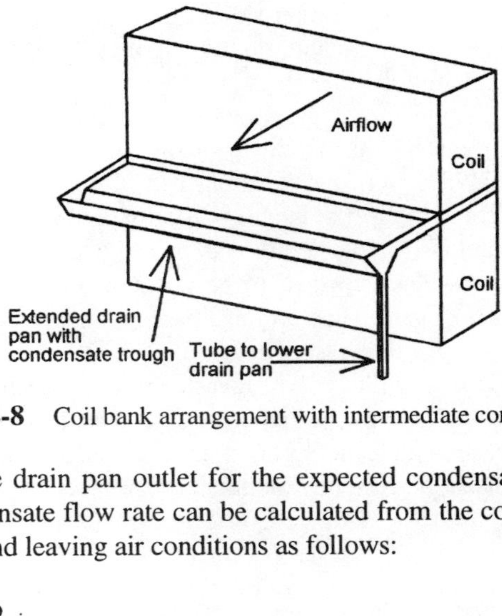

FIGURE 4-8 Coil bank arrangement with intermediate condensate pan.

Size the drain pan outlet for the expected condensate volume. The condensate flow rate can be calculated from the cooling coil's entering and leaving air conditions as follows:

$$Q_{cond} = \frac{\rho_{air}}{\rho_{water}} Q_{air}(W_{ec} - W_{lc}) = C_{14} Q_{air}(W_{ec} - W_{lc}) \qquad (4\text{-}53)$$

where

Q_{cond} = condensate flow rate, gpm (L/s)

Q_{air} = airflow rate, cfm (L/s)

ρ_{air} = air density, 0.075 lb/ft^3 (1.2 kg/m^3)

ρ_{water} = water density, 8.33 lb/gal (1000 kg/m^3)

W_{ec} = humidity ratio entering coil, lb_w/lb_a (kg_w/kg_a)

W_{lc} = humidity ratio leaving coil, lb_w/lb_a (kg_w/kg_a)

C_{14} = 0.009 $lb_w \cdot gal/lb_a \cdot ft^3$ (0.0012 $kg_w \cdot L/kg_a \cdot m^3$)

For the systems described in the examples of sections 4.10.8 and 4.10.9, the condensate flow is calculated as follows.

For a conventional design:

Q_{cond} = 0.009 × 21,396 (0.0100 – 0.0087) = 0.250 gpm

(Q_{cond} = 0.0012 × 10, 092 (0.0100 – 0.0087) = 0.016 L/s)

For a cold-air design:

Q_{cond} = 0.009 × 13, 430 (0.0088 – 0.0062) = 0.314 gpm

(Q_{cond} = 0.0012 × 6, 278 (0.0088 – 0.0062) = 0.020 L/s)

The drain pan outlet should be on the downstream side of the coil. The condensate drain trap leg must be at least long enough so the column of water in the trap will overcome the fan pressure in the coil plenum. If the trap leg is too short on a negative-pressure drain, water will not flow out the drain but will accumulate and eventually overflow the drain pan. On a positive-pressure drain, insufficient trap leg length will allow air to flow out of the drain continuously. Note that the fan pressure used for sizing a trap is the highest value expected to be encountered. For negative-pressure traps, this value is determined based on a dirty filter with the maximum expected pressure drop.

Figure 4-9 illustrates suggested sizing guidelines for drain traps. If the horizontal drain pipe is longer than 15 ft (4.6 m), replace the downstream cleanout plug with a screened cover so the horizontal drain can be designed for gravity flow.

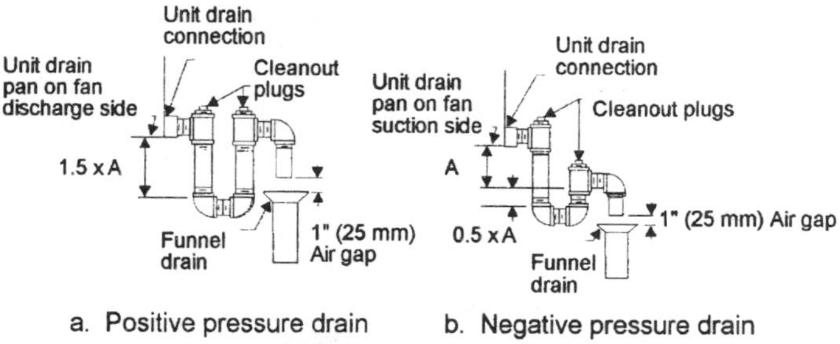

Note: "A" = System static pressure in inches (static pressure in Pa x 0.0124 mm), at drain point.
Condensate drain pipe size = unit drain connection size (minimum 1" (25 mm))

FIGURE 4-9 Condensate drain sizing.

4.11.9 MAINTENANCE

Coil maintenance considerations include air filtration and cleaning of coils and condensate drains. The increased flow of condensate across coils in cold-air distribution systems may have a cleaning effect, reducing the amount of dust and dirt that lodges in the coils. However, buildings in many climates experience extended periods when coils perform little or no dehumidification.

Coil maintenance should emphasize frequent filter inspection and service. Reducing the amount of dust reaching the coils is preferable to cleaning coils that have become plugged. Designers should consider specifying high-efficiency filters to provide increased coil protection. Many designers specify 85% efficient filters with 30% efficient prefilters, as rated by the ASHRAE dust spot method (ASHRAE 1992). The use of these high-efficiency filters can substantially reduce coil-cleaning requirements, although the increased pressure drop does represent an increase in fan energy consumption. This improved filtration also offsets any reduction in particulate removal resulting from the lower total volume of supply air circulating through the space.

In general, coils should be cleaned at least annually if high-efficiency filtration is not used. Specifying double-wall construction on air-handling units allows the use of pressurized water for cleaning coils without damaging insulation.

Special care is required at start-up to prevent construction dust from being deposited on wet coil surfaces. Remove as much dust as possible from the conditioned space and from the duct system before cooling coils are started up.

Condensate drains should be inspected regularly and cleaned if necessary. The increased volume of condensate may provide a beneficial flushing effect. However, if drains do become plugged, drain pans will quickly overflow and the increased condensate flow will create an increased cleanup problem.

4.11.10 LIFE-CYCLE COSTS

The preferred cooling coil selection is the one that minimizes the life-cycle cost of installing and operating the coil. First costs are most dependent on the number of rows and the face area. The face area often influences the cost of the air-handling unit enclosure, as well as that of the coil itself. Operating costs include pumping energy, determined by the fluid pressure drop, and fan energy, determined by the air pressure drop.

Minimizing the number of rows generally reduces the coil first cost, as well as air and fluid pressure drops and their associated operating costs. Increasing the face area reduces the air pressure drop. In some cases, a coil with a larger face area can provide the required heat transfer performance with fewer rows, at approximately the same first cost. However, the larger coil may require a larger air-handler cabinet at a higher cost.

Life-cycle cost analyses tend to favor larger coils and lower face velocities than indicated by traditional rules of thumb. Wheeler (1977) discusses life-cycle cost analysis for cooling coils and recommends 300- to 400-fpm coil face velocity. Tamblyn (1992) presents an extensive analysis of coil selections for 45°F (7.2°C) supply air and finds that lower face velocities and wider fluid temperature ranges are often recommended.

4.11.11 COOLING COIL SELECTION EXAMPLES

The following examples illustrate cooling coil selections for three air distribution options:

- typical conventional selection with 55°F (12.8°C) leaving air and 12°F (6.6°C) fluid temperature rise,
- cold-air distribution selection with 45°F (7.2°C) leaving air and 16°F (8.8°C) fluid temperature rise, and
- cold-air distribution selection with 42°F (5.6°C) leaving air and 24°F (13.2°C) fluid temperature rise.

Coils are selected for the prototype system presented in previous examples. The entering air conditions are determined for each option by the methods given in section 4.10. The selection parameters are summarized in Table 4-11.

Supplying these parameters to a manufacturer's coil selection program yields a number of possible selections. The results for each option are discussed below.

55°F (12.8°C) Supply Air, 12°F (6.6°C) Range

Table 4-12 shows coil selections for a typical conventional air distribution design. The four-row coil with 11.6 fins/in. (0.46 fins/mm) and no internal tube enhancement was selected as a reference for comparing costs of the other selections. The cost comparisons in each of the examples are given in terms of the ratio of each coil's cost to that of this reference selection, based on budget figures provided by a supplier.

TABLE 4-11 PARAMETERS FOR COIL SELECTION EXAMPLES

	Leaving air temperature, °F (°C)		
	55 (12.8)	**45 (7.2)**	**42 (5.6)**
Fluid temp. rise, °F (°C)	12 (6.6)	16 (8.8)	24 (13.2)
Supply air volume, cfm (L/s)	20,400 (9627)	13,300 (6276)	12,100 (5709)
Entering air dry-bulb, °F (°C)	77 (25.0)	79 (26.1)	79 (26.1)
Entering air wet-bulb, °F (°C)	64 (17.8)	63 (17.2)	63 (17.2)
Fluid	Water	Water	Water or 25% Ethylene Glycol
Entering fluid temp., °F (°C)	44 (6.7)	37 (2.8)	36 (2.2)
Face velocity, fpm (m/s)	450-550 (2.3-2.8)	350-500 (1.8-2.5)	350-500 (1.8-2.5)
Max. air Δp, in. w.g. (Pa)	1.0 (249)	1.0 (249)	1.0 (249)
Max. fluid Δp, ft. (m)	20 (6.1)	25 (7.6)	25 (7.6)

TABLE 4-12 55°F (12.8°C) COIL SELECTIONS

Rows	Fins per		Air Pressure Drop		Water Pressure Drop		Internal Tube Enhancement	Relative Cost
	in.	mm	in. w.g.	Pa	ft H_2O	kPa		
Face Velocity: 504 fpm (2.6 m/s). Face Area: 40.5 ft² (3.8 m²)								
4	9.4	0.37	0.57	142	11.6	34.8	Yes	0.97
4	11.6	0.46	0.64	159	4.3	12.8	No	1.00
6	6.7	0.26	0.70	174	6.3	18.7	No	1.36

45°F (7.2°C) Supply Air, 16°F (8.8°C)Range

Table 4-13 summarizes the coil selections for 45°F (7.2°C) supply air with a 16°F (8.8°C) fluid temperature rise. Table 4-13 shows that many coil options are available to provide the required performance.

The selections include coils in three sizes, resulting in three options for face velocity. Coils with 6, 8, or 10 rows could be used, with fin spacing ranging from 7 to 13 fins per inch (0.27 to 0.51 fins per mm).

The use of turbulence enhancement in the coil tubes to increase water-side heat transfer allows coils to provide the required performance with fewer rows and fins. This reduces air pressure drop but increases water pressure drop. In this case, the costs of similar coils with and without internal tube enhancement are essentially the same.

TABLE 4-13 45°F (7.2°C) COIL SELECTIONS

Rows	Fins per		Air Pressure Drop		Water Pressure Drop		Internal Tube Enhancement	Relative Cost
	in.	mm	in. w.g.	Pa	ft.H_20	kPa		
Face Velocity: 316 fpm (1.6 m/s). Face Area: 40.5 ft^2 (3.8 m^2)								
6	9.5	0.37	0.41	102	11.0	32.8	Yes	1.35
6	13.2	0.52	0.53	132	4.1	12.2	No	1.34
8	6.8	0.27	0.43	107	14.4	43.0	Yes	1.72
8	8.1	0.32	0.49	122	5.3	15.8	No	1.75
Face Velocity: 366 fpm (1.9 m/s). Face Area: 35.0 ft^2 (3.3 m^2)								
6	10.5	0.41	0.55	137	12.9	38.6	Yes	1.27
8	7.3	0.29	0.59	147	17.0	50.8	Yes	1.56
8	8.8	0.35	0.66	164	6.2	18.4	No	1.51
Face Velocity: 418 fpm (2.1 m/s). Face Area: 30.6 ft^2 (2.9 m^2)								
6	11.5	0.45	0.73	182	16.2	48.5	Yes	1.21
8	7.9	0.31	0.78	194	21.4	63.8	Yes	1.33
8	9.6	0.38	0.86	214	7.7	23.1	No	1.36
10	7.1	0.28	0.89	221	9.5	28.5	No	1.69

The six-row coils are least expensive, and the coils with lower face velocities minimize the air pressure drop. The final selection would normally be based on an analysis of the overall life-cycle cost of each coil.

42°F (5.6°C) Supply Air, 24°F (13.2°C) Range

Table 4-14 summarizes the coil selections for 42°F (6°C) supply air with a 24°F (13°C) water temperature rise, with chilled water. Table 4-15 lists the selections for 25% ethylene glycol solution.

The lower supply air temperature and increased fluid temperature range result in selections with more rows and fins, and lower face velocities, than for the previous example. A comparison of Tables 4-14 and 4-15 shows that the use of ethylene glycol heat transfer fluid results in a 10% to 15% increase in fin density, a 10% increase in air pressure drop, and a 25% increase in fluid pressure drop. The coils selected for ethylene glycol are slightly more expensive than the coils selected for water.

Several possible selections of double-circuit coils also could meet the requirements specified for this example. These coils, in which only half of the full fluid volume flows through each coil row, are nor-

TABLE 4-14 42°F (5.6°C) COIL SELECTIONS, WATER

Rows	Fins per		Air Pressure Drop		Water Pressure Drop		Internal Tube Enhancement	Relative Cost
	in.	mm	in. w.g.	Pa	ft. H_20	kPa		
Face Velocity: 283 fpm (1.4 m/s). Face Area: 40.5 ft² (3.8 m²)								
8	12.8	0.50	0.56	139	7.3	21.8	Yes	1.81
10	9.2	0.36	0.55	137	9.0	26.9	Yes	2.00
12	7.3	0.29	0.55	137	10.7	32.0	Yes	2.36
12	10.6	0.42	0.72	179	3.8	11.4	No	2.38
Face Velocity: 329 fpm (1.7 m/s). Face Area: 35.0 ft² (3.3 m²)								
10	10.1	0.40	0.75	187	10.7	32.0	Yes	1.88
12	7.9	0.31	0.77	192	12.7	38.1	Yes	2.19
12	11.4	0.45	0.97	241	4.5	13.6	No	2.25
Face Velocity: 376 fpm (1.9 m/s). Face Area: 30.6 ft² (2.8 m²)								
10	11.0	0.43	0.98	244	13.4	40.1	Yes	1.75
12	8.5	0.33	1.01	251	16.0	47.8	Yes	2.03

TABLE 4-15 42°F (5.6°C) COIL SELECTIONS, 25% ETHYLENE GLYCOL

Rows	Fins per		Air Pressure Drop		Water Pressure Drop		Internal Tube Enhancement	Relative Cost
	in.	mm	in. w.g.	Pa	ft. H_20	kPa		
Face Velocity: 283 fpm (1.4 m/s). Face Area: 40.5 ft² (3.8 m²)								
10	10.4	0.41	0.60	149	11.3	33.8	Yes	2.04
12	8.0	0.31	0.60	149	13.4	40.2	Yes	2.33
Face Velocity: 329 fpm (1.7 m/s). Face Area: 35.0 ft² (3.3 m²)								
10	11.6	0.46	0.82	204	13.7	41.0	Yes	1.83
12	8.8	0.34	0.82	204	16.4	48.9	Yes	2.22
Face Velocity: 376 fpm (1.9 m/s). Face Area: 30.6 ft² (2.8 m²)								
10	12.8	0.51	1.12	279	17.6	52.7	Yes	1.78
12	9.5	0.37	1.07	266	21.1	63.0	Yes	2.06

mally used in high-flow, low delta-T applications. While they provide the full-load performance required for this example with a low fluid-side pressure drop, double-circuit coils are not recommended for applications where fluid flow to the coils varies with the load.

4.12 REFERENCES

ARI. 1991. *ARI Standard 410-91, Forced-circulation air-cooling and air-heating coils*. Arlington, Va.: Air-Conditioning and Refrigeration Institute.

ASHRAE. 1989. *ASHRAE Standard 90.1-1989, Energy-efficient design of new buildings except low-rise residential buildings*. Atlanta: American Society of Heating, Refrigerating and Air-Conditioning Engineers, Inc.

ASHRAE. 1992. *ANSI/ASHRAE Standard 52.1, Gravimetric and dust-spot procedures for testing air-cleaning devices used in general ventilation for removing particulate matter*. Atlanta: American Society of Heating, Refrigerating and Air-Conditioning Engineers, Inc.

ASHRAE. 1993. *1993 ASHRAE handbook—Fundamentals*. Atlanta: American Society of Heating, Refrigerating and Air-Conditioning Engineers, Inc.

ASHRAE. 1994a. *ANSI/ASHRAE Standard 15-1994, Safety code for mechanical refrigeration*. Atlanta: American Society of Heating, Refrigerating and Air-Conditioning Engineers, Inc.

ASHRAE. 1994b. *1994 ASHRAE handbook—Refrigeration*. Atlanta: American Society of Heating, Refrigerating and Air-Conditioning Engineers, Inc.

ASHRAE. 1996 *ASHRAE handbook—HVAC systems and equipment*. Atlanta: American Society of Heating, Refrigerating and Air-Conditioning Engineers, Inc.

ASTM. 1980. *ASTM Standard E96-80, Standard test methods for water vapor transmission of materials*. Philadelphia: American Society for Testing and Materials.

Brooks, P.J. 1995. Duct design fundamentals. *ASHRAE Journal* 37(4): 69-76.

Carpenter, J.H. 1984. *Performance and control of fans in variable air volume systems*. Syracuse, N.Y.: Carrier Corporation.

Colliver, D.G., R.S. Gates, H.Z. Zhang, and K.T. Priddy. 1995. Determination of the 1%, 2.5%, and 5% occurrences of extreme dew-point temperatures and mean coincident dry-bulb temperatures. *ASHRAE Transactions* 101(1).

Dorgan, C.E., and C.B. Dorgan. 1995. *Cold air distribution design guide*. 2nd ed. EPRI TR-105604. Palo Alto, Calif.: Electric Power Research Institute.

Dorgan, C.E., and J.S. Elleson. 1987. *Field evaluation of cold air distribution systems*. EPRI EM-5447. Palo Alto, Calif.: Electric Power Research Institute.

Dorgan, C.E., and J.S. Elleson. 1988. *Cold air distribution design guide*, 1st ed. EPRI EM-5730. Palo Alto, Calif.: Electric Power Research Institute.

Dorgan, C.E., and J.S. Elleson. 1993. *Design guide for cool thermal storage*. Atlanta: American Society of Heating, Refrigerating and Air-Conditioning Engineers, Inc.

Fields, W.G. 1989. Operation and cost advantages of a liquid overfeed low-temperature air distribution system. *ASHRAE Transactions* 95(1): 1300-1307.

Gatley, D.P. 1992. *Cool storage ethylene glycol design guide*. EPRI TR-100945. Palo Alto, Calif.: Electric Power Research Institute.

Harmon, J.J., and H.C. Yu. 1993. Cold air distribution and concerns about condensation. *ASHRAE Journal* 35(5): 34-40.

Lorsch, H.G., et al. 1993. *Air-conditioning systems design manual*. Atlanta: American Society of Heating, Refrigerating and Air-Conditioning Engineers, Inc.

McQuiston, F.C., and J.D. Spitler. 1992. *Cooling and heating load calculation manual*, 2d ed. Atlanta: American Society of Heating, Refrigerating and Air-Conditioning Engineers, Inc.

Mirth, D.R., S. Ramadhyani, and D.C. Hittle. 1993. Thermal performance of chilled-water cooling coils operating at low water velocities. *ASHRAE Transactions* 99(1): 43-53.

Nussbaum, O.J. 1990. Using glycol in a closed circuit system. *Heating/Piping/Air Conditioning,* January, pp. 75-85.

Scofield, M., and G. Fields. 1989. Joining VAV and direct refrigeration. *Heating/Piping/Air Conditioning*, September, pp. 137-152.

SMACNA. 1985. *HVAC air duct leakage test manual*, 1st ed. Vienna, Va.: Sheet Metal and Air Conditioning Contractors' National Association, Inc.

SMACNA. 1990. *HVAC systems duct design*, 3d ed. Vienna, Va.: Sheet Metal and Air Conditioning Contractors' National Association, Inc.

Stewart, W.E., C.K. Saunders, and L.A. Stickler. 1994. Latent leakage. Final report of ASHRAE Research Project RP-746. Atlanta: American Society of Heating, Refrigerating and Air-Conditioning Engineers, Inc.

Stewart, W.E., C.K. Saunders, and L.A. Stickler. 1995. Air pressure control of latent leakage. *ASHRAE Journal* 37(5): 40-47.

Tamblyn, R.T. 1992. Common sense cooling coils. *Heating/Piping/Air Conditioning*, December, pp. 49-57.

Tsal, R.J., H.J. Behls, and R. Mangel. 1988. T-method duct design, part I: Optimization theory, part II: Calculation procedure and economic analysis. *ASHRAE Transactions* 94(2): 90-111.

Wheeler, A.E. 1977. Air handling unit design for energy conservation. *ASHRAE Journal* 19(6): 47-52.

CHAPTER FIVE

SELECTION OF COLD-AIR DIFFUSION COMPONENTS

5.1 INTRODUCTION

An acceptable thermal environment is achieved through proper selection of cold-air diffusion components. The main cold-air diffusion components to be chosen in a cold-air system are the diffusers and the mixing boxes. In this chapter, information about selection and sizing of diffusers and mixing boxes for cold-air systems is given. The chapter provides a description of the different types of diffusers used to produce cold-air jets followed by a introductory model of jet mixing. Next are guidelines for selecting various candidate cold-air diffusers. The selection of mixing boxes for cold-air systems is also addressed.

Proper selection of cold-air diffusers is accomplished by consideration of the primary characteristics of the cold-air jet. Jet diffusion and mixing is a complex topic in fluid mechanics. Simple models of jet mixing predict the mixing and entrainment of jets. These models have been incorporated into design procedures for diffuser selection. In this chapter a step-by-step design example is given to show how the current diffuser selection procedure given in chapter 31 of the *1993 ASHRAE Handbook—Fundamentals* (ASHRAE 1993) is extended to cold-air systems.

The three main design parameters for any type of diffuser, including cold-air diffusers, are the room characteristic length, the throw, and the separation distance. These design parameters are discussed in sections 5.3, 5.4, and 5.5. For use in diffuser selection, the length, throw, and separation design parameters are cast into nondimensional ratios, such as the ratio of the throw to the room characteristic length. The design procedure detailed in this chapter uses optimum values of these ratios for diffuser selection. Design examples are given that illustrate the determination of the throw and the separation distance of a cold-air jet based on the diffuser size, type, and discharge conditions.

Another important parameter in jet diffusion is the entrainment ratio, the ratio of the jet flow volume at a given distance from the outlet to the outlet jet flow volume. Proper selection of the throw and separation distance, as outlined in section 5.7, will lead to sufficient entrainment of the room air by the diffuser jet.

5.2 DIFFUSERS FOR COLD-AIR DISTRIBUTION

5.2.1 DIFFUSER TYPES

There are a number of diffuser types available for cold-air systems. Some types are conventional diffusers validated for cold air, others are specially designed for cold-air systems. If chosen properly, both conventional and cold-air diffusers will operate satisfactorily in a cold-air distribution system. The primary function of a diffuser is to provide the required amount of outlet flow momentum to achieve satisfactory mixing in the room being conditioned. This airflow momentum depends on proper selection of the outlet diffuser size.

The main difference between conventional and cold-air diffusers is that the cold-air diffusers must operate satisfactorily over a wider range of temperatures and flow rates than conventional diffusers. With this wider range of temperatures and flow rates, considerations such as adequate room air mixing, jet separation, and noise must be addressed.

Generally speaking, diffusers can be grouped into two major categories: linear and radial. Linear diffusers produce a jet from a line source, and radial diffusers produce a jet originating from a point source. Types of linear diffusers include slot and jet diffusers. Types of radial diffusers include circular or round, square, crossflow, swirl, and perforated diffusers. Additional information about different diffuser types is given in chapter 17, "Air-Diffusing Equipment," in the *1996 ASHRAE Handbook—HVAC Systems and Equipment* (ASHRAE 1996).

The most commonly used diffuser types are slot and circular diffusers. Jet diffusers have also been recommended by a number of manufacturers for cold-air distribution applications, due to the relatively high momentum that they impart to the room air.

Slot Diffuser

The slot diffuser is a ceiling-mounted diffuser that produces a linear air jet from a line source. The exit airflow can be configured to produce one-way or two-way throws. Slot diffusers contain an internal curved blade located at the exit. This blade causes the airflow to deflect toward and adhere to the ceiling. Slot diffusers are typically 24 in. to 60 in. (609.6 mm to 1524.0 mm) long, with slot widths varying from 1/2 in. to 1 in. (12.7 mm to 24.4 mm), and can have from one to eight slots. Typical cross sections and flow profiles of one- and two-way slot diffusers are shown in Figure 5-1. The sideways or lateral spread of the jet from a slot diffuser is very small in the horizontal direction compared to the vertical or downward spread.

Radial Diffuser

The radial diffuser is a ceiling-mounted diffuser that produces a radial jet. A radial jet has a point as an origin, in contrast to the linear diffuser type, which has a line as the jet origin. Radial diffusers

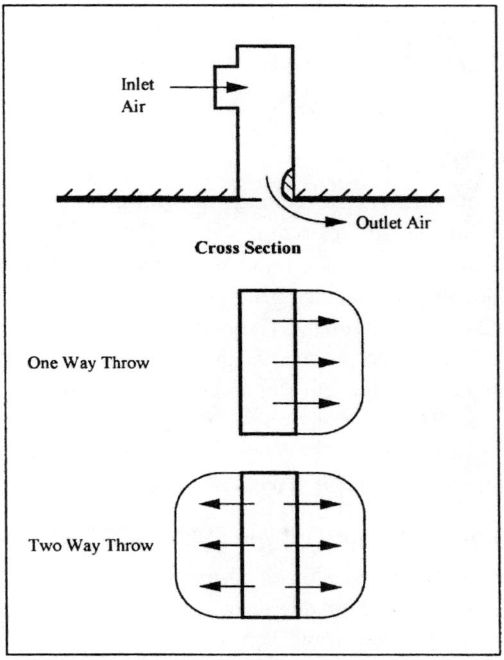

FIGURE 5-1 (Top view) Slot diffuser flow.

are composed of concentric air passages that can be round, square, or rectangular. The round opening produces a circular airflow. The internal geometry can be arranged to produce either a radial or a cross-flow pattern. Typical cross sections and flow profiles of round and square radial diffusers are shown in Figure 5-2.

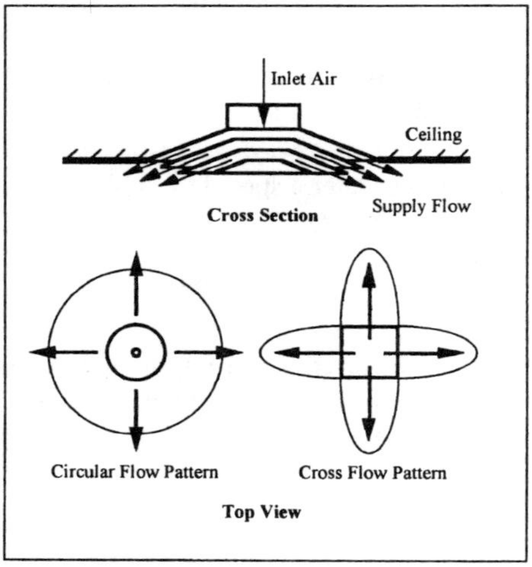

FIGURE 5-2 Radial diffuser flow profile.

Jet Diffuser

Jet diffusers are ceiling- or wall-mounted diffusers that have multiple small area outlets. The outlets produce multiple jets from either a linear or radial source. There is a range of outlet areas available, with the nozzle openings varying from 1/2 in. to 1 in. (1.3 cm to 24.4 mm) diameter. The multiple jets merge at 5 to 10 diameters downstream to produce a single jet. The outlet areas are generally circular in shape, but slot and triangular outlets also have been produced. The cross section and flow profile of a linear jet diffuser are shown in Figure 5-3. With a radial jet diffuser, the multiple small outlets are arranged in a circular fashion.

Additional Diffuser Types: Swirl and Perforated

Swirl diffusers are ceiling-mounted diffusers that produce an outlet flow with a high degree of spin about an axis parallel to the direction of the main flow. The spin is produced by internal blades that impart a rotational velocity to the flow as it exits the diffuser. They have a relatively high entrainment due to the swirl of the exit air.

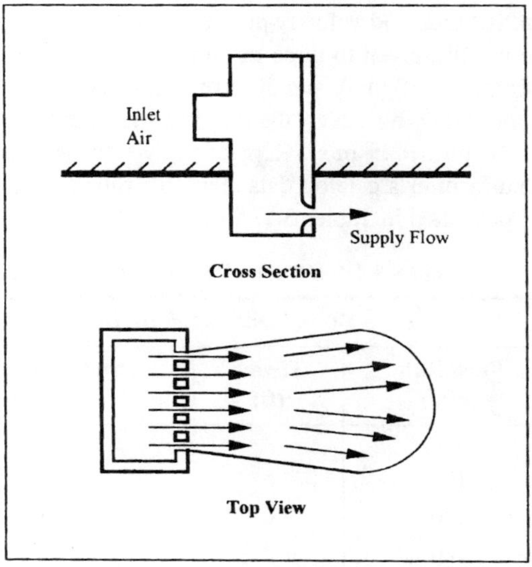

FIGURE 5-3 Jet diffuser flow profile.

Perforated ceiling diffusers are formed from perforated plates mounted flush with the ceiling. Since the initial outlet jets are vertical, perforated plate diffusers are more susceptible to separation. They can have internal louvers that will also add some swirl velocity to the outlet flow. The open area is from 10% to 50%, which determines the capability of the diffuser.

Swirl and perforated diffusers currently are not used to a large extent for cold-air distribution. As long as they meet throw and separation distance requirements, they would also be suitable candidates as cold-air diffusion components.

5.2.2 REPRESENTATIVE CATALOG DATA

Manufacturers' catalog data for diffusers list performance information in tabular form as a function of the diffuser airflow rate. The performance information includes the throw distance, the static pressure drop, and the noise level:

- T = throw distance to terminal velocity, ft (m)
- ΔP = static pressure drop, in. w.g. (Pa)
- NC= noise criterion

The diffuser performance is obtained in accordance with ASHRAE standards for diffuser testing. The presentation format will vary from manufacturer to manufacturer. Some manufacturers will list total and velocity pressures for the diffuser. The throw data are usually given to three terminal velocities: 150 fpm (0.75 m/s), 100 fpm (0.50 m/s), and 50 fpm (0.25 m/s). For the data presented in the following tables, the throw distance is to a terminal velocity of 50 fpm (0.25 m/s). Representative diffuser data taken from a manufacturer's catalog data sheet for round, slot, and jet diffusers are presented in Tables 5-1, 5-2, and 5-3.

TABLE 5-1 ROUND DIFFUSER CATALOG DATA

Round Diffuser, 6-in. Dia. Inlet			
Flow Rate (cfm)	**Throw (ft)**	**Static Pressure (in. w.g.)**	**NC**
80	4	0.016	10
100	5	0.025	10
120	6	0.036	14
140	7	0.048	19
160	8	0.062	23
180	9	0.079	26
200	10	0.098	30
235	11	0.140	35
275	13	0.189	40

TABLE 5-2 SLOT DIFFUSER CATALOG DATA

Slot Diffuser 1-in. slot, 1 slot, 48-in. Long			
Flow Rate (cfm)	**Throw (ft)**	**Static Pressure (in. w.g.)**	**NC**
68	12	0.019	10
93	16	0.035	20
117	21	0.058	23
141	23	0.081	26
166	25	0.111	28
190	26	0.147	30
215	28	0.186	32
239	30	0.232	34
264	31	0.281	35

TABLE 5-3 JET DIFFUSER CATALOG DATA

Jet Diffuser, 1 way, 48-in. Long			
Flow Rate (cfm)	**Throw (ft)**	**Static Pressure (in. w.g.)**	**NC**
40	8	0.012	10
60	17	0.026	16
80	19	0.047	21
100	22	0.073	26
120	24	0.105	29
140	26	0.143	33
160	28	0.187	37
180	29	0.237	40
200	31	0.292	43
220	32	0.354	45
240	34	0.421	48
260	35	0.494	50
280	37	0.573	53

5.3 ROOM CHARACTERISTIC LENGTH AND DIFFUSER PLACEMENT

The room characteristic length, *L*, depends on the type of diffuser, the diffuser location, and the distance to a wall or symmetry plane. The room characteristic length represents the horizontal length in the room to which the outlet cold-air jet needs to extend. It is usually equal to some fraction of the wall-wall distance. The characteristic room lengths for linear and radial diffusers are defined in Table 5-4.

TABLE 5-4 ROOM CHARACTERISTIC LENGTH, *L*

Diffuser Type	Characteristic Length
Linear diffuser	Distance to wall or mid-plane between outlets
Radial diffuser	Distance to closest wall or intersecting air jet

The room characteristic length is found from the diffuser layout pattern. The diffuser layout is determined by subdividing a plan view of the room into squares or rectangles. Circular diffusers are located at the center of the square subdivisions, and linear diffusers are located along the middle or edge of the rectangular subdivisions. Linear diffusers, such as jet and slot diffusers, are typically located in line with each other, not blowing at each other. If one-way throw is used, they are located at an edge near a wall, and if two-way throw is used, they are located along a centerline of the rectangular subdivision, as shown in Figure 5-4.

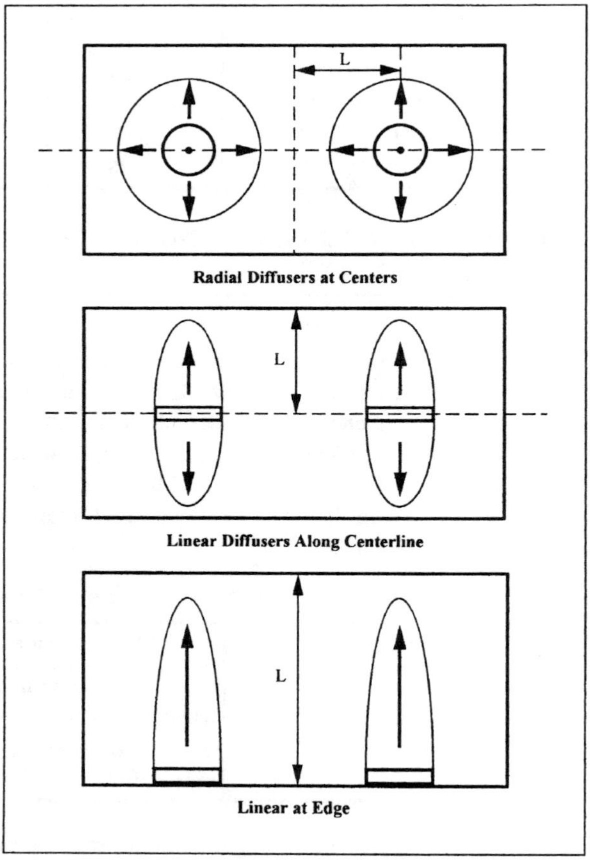

FIGURE 5-4 Room characteristic lengths for various diffusers.

5.4 COLD-AIR JET THROW

The outlet momentum of the cold-air jet is transferred to the room air through unsteady turbulent diffusion. This diffusion will cause the jet velocity to decrease. The length that the jet travels during this diffusion process is termed the *throw* of the jet. The throw, *T*, of an air jet from a diffuser is the distance from the diffuser outlet to the point where the supply jet centerline velocity has decreased to a specified terminal velocity. A schematic of the throw distance is shown in Figure 5-5 for a ceiling-mounted diffuser. The terminal velocity is usually selected to be 50 fpm (0.25 m/s), but 100 fpm (0.50 m/s) has also been used for ceiling slot diffusers.

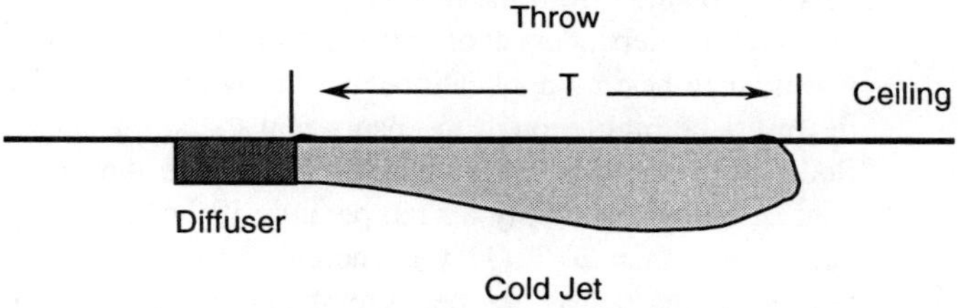

FIGURE 5-5 Cold air jet throw.

The throw of a diffuser is a significant design variable since it has been found to be a reliable indicator of acceptable room air mixing. The throw is proportional to the momentum of the outlet supply air, so it is a direct measure of the "stirring" and entrainment of the room air by the jet. There are two limits to acceptable throw values. If the throw is too short, relative to the room length, the room air will tend to be stagnant and stratified, with poor mixing. If the throw is too long, the room will be drafty. Diffusers used for both heating and cooling should have sufficient throw to properly mix the room air during both heating and cooling operation.

Tests of diffusers over a wide range of loads, flow rates, and temperature differences have resulted in recommendations for *T/L*, the ratio of diffuser throw, *T*, to room characteristic length, *L*, that result in a maximum value of the air diffusion performance index (ADPI), discussed in section 3.5. The existing throw guidelines, obtained from conventional air supply tests, that result in acceptable room air motion, can also be used for cold-air jets (Miller 1991). The recommended throw values depend on the type of diffuser and the room load and are based on maximizing the room thermal comfort.

The tests also recommend a range of values of T/L where the ADPI is above a minimum 80% value. This range generally extends over a factor of 6—0.5 to 3.3—for slot diffusers and a factor of 2 for round or circular diffusers. Therefore, ceiling-mounted linear slot diffusers will perform well over a wide range of flow rates.

The dependence of the throw on the supply air temperature depends on the type of diffuser. Ceiling diffusers generally show a small effect of less than 10% (Hassani et al. 1993) since the basic jet flow pattern is not changed. However, if the outlet flow pattern from the diffuser changes with a decrease in the supply air temperature, the throw will be affected. It is recommended that cold-air throw data be obtained from the diffuser manufacturer as a check on the temperature dependence of the diffuser throw. When delivering warm air in heating applications, the throw and outlet flow velocities must be high enough to ensure that the warm air reaches the floor and to reduce thermal stratification. The throw is generally unaffected by the supply air temperature if the temperature difference is less than 20°F (11°C). Therefore, to prevent stratification, the diffuser outlet air temperature should not be more than 20°F (11°C) above the occupied zone air temperature (ASHRAE 1993, chapter 31).

The recommended throw ratios for cooling applications are given in Table 5-5 (ASHRAE 1993, chapter 31). For example, the ratio of diffuser throw, T, to room characteristic length, L, recommended for ceiling slot diffusers is from 0.3 to 1.5, and the ratio of diffuser throw, T, to room characteristic length, L, recommended for circular ceiling diffusers is from 0.7 to 1.5., depending on the room load.

5.5 COLD-AIR JET SEPARATION

The separation distance is the distance from the outlet to the point where the supply jet separates from the ceiling. Acceptable separation distances are those longer than the room characteristic length. With the advent of colder supply temperatures, the potential for jet separation has increased. A supply jet should diffuse and mix along the ceiling above the occupied zone. However, at cold air temperatures, it is possible for the cold-air jet to separate from the ceiling before adequate mixing has occurred, as shown in Figure 5-6.

TABLE 5-5 AIR DIFFUSION PERFORMANCE (ADPI) SELECTION GUIDE

Diffuser Type	Terminal Velocity fpm (m/s)	Room Load Btuh/ft^2 (W/m^2)	*T/L* for Max. ADPI	Maximum ADPI	For ADPI Greater Than	Range of *T/L*
Slot	100 (0.50)	80 (252)	0.3	85	80	0.3 - 0.7
		60 (189)	0.3	88	80	0.3 - 0.8
		40 (126)	0.3	91	80	0.3 - 1.1
		20 (63)	0.3	92	80	0.3 - 1.5
Slot	50 (0.25)	40 (126)	1.0	91	80	0.5 - 3.3
		20 (63)	1.0	91	80	0.5 - 3.3
Round	50 (0.25)	80 (252)	0.8	76	70	0.7 - 1.3
		60 (189)	0.8	83	80	0.7 - 1.2
		40 (126)	0.8	88	80	0.7 - 1.5
		20 (63)	0.8	93	90	0.7 - 1.3
Perforated	50 (0.25)	35-360 (110-1136)	2.0	96	90 80	1.4 - 2.7 1.0 - 3.4

With separation, a cold air draft will enter the occupied space, resulting in unacceptable thermal discomfort. The thermal discomfort is due to two factors—the cold draft of the separated jet in the occupied space and the lack of adequate mixing in the stagnant areas of the room not reached by the separated jet. The relative extent of these two factors and the resulting effect on the ADPI are discussed further in Kirkpatrick and Knappmiller (1996).

The separation distance parameter, x_S, is the distance from the diffuser at which a jet separates from the ceiling. Jets from ceiling diffusers initially have a tendency to attach to the ceiling surface, due to their low pressure relative to the room. However, cold-air jets will detach from the ceiling if the downward buoyancy force result-

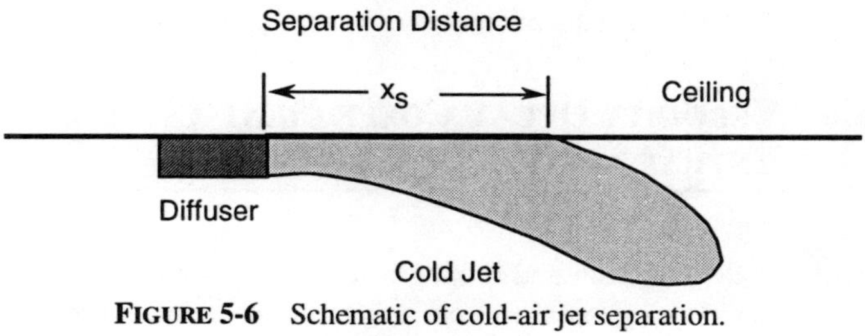

FIGURE 5-6 Schematic of cold-air jet separation.

ing from the temperature difference between the jet and the ambient room conditions is larger than the force exerted by the ambient pressure acting upward and holding the jet next to the ceiling.

Determination of the separation distance has been the focus of a number of research activities (Hassani et al. 1993). The research results indicate that the separation distance can be correlated with the outlet jet conditions. The separation distance depends on the outlet conditions of the jet, including the velocity decay constant K (discussed in section 5-6), the outlet temperature, flow rate, and static pressure drop. An equation that has been used to correlate data from slot and round diffusers is:

$$x_s = aC_s K^{1/2}\left(\frac{\Delta T}{T}\right)^{-1/2} Q^{1/4} \Delta P^{3/8} \tag{5-1}$$

where

x_S =jet separation distance, ft (m)
C_S =separation coefficient, $C_S = 1.2$
K =velocity decay coefficient, dimensionless
a =constant, $a = 11.91$ (48.04)
ΔT =room-jet temperature difference, °F (°C)
T =average absolute room temperature, °R (K)
ΔP =diffuser static pressure drop, in. w.g. (Pa).

A representative value of the separation coefficient, C_S, that has been found to best match the results of analyses and experiments of a wide variety of diffusers is 1.2 (Rodhal 1977; Nielsen 1987; Hassani et al. 1993). Additional discussion of the separation coefficient and its sensitivity to room airflow conditions is contained in Knappmiller and Kirkpatrick (1994).

The separation distance for the cold-air jets modeled with Equation 5-1 is linearly proportional to flow rate. This can be seen by inspection of Equation 5-1. Since the pressure drop across a diffuser is proportional to the square of the flow rate, the pressure drop to the 3/8 power is equivalent to a flow rate to the 3/4 power.

5.6 VELOCITY DECAY COEFFICIENT AND JET ENTRAINMENT

Jet diffusion and mixing is a complex topic in fluid mechanics. A cold air jet issuing from a diffuser is at a lower pressure than the room air and entrains the room air through a turbulent mixing process. The jet contains unsteady turbulent eddies that sweep room air into the main body of the jet. As the jet mixes with the room air, its momentum is transferred to the room air, its velocity decreases, and it becomes larger. The energy exchange due to mixing also causes the room air temperature to decrease as the cold-air jet temperature increases.

Simple models of jet mixing are used to predict the entrainment of cold air jets. The decrease in the velocity of the cold air jet in what is termed the "third zone" (chapter 31, "Space Air Diffusion," of the *1993 ASHRAE Handbook—Fundamentals*) is modeled with Equation 5-2. A similar equation is used for the temperature profile. The model predicts the velocity and temperature change along the streamwise direction of the jet as the room air is entrained. This equation has been shown to be applicable to a wide variety of jets from different types of diffusers. The equation uses a velocity decay coefficient, K, to determine the jet centerline velocity at a distance x from the diffuser as a function of the initial outlet velocity and effective outlet area. The effective outlet area is not the physical outlet area but is found from pressure drop and flow rate considerations.

$$\frac{V_x}{V_o} = K\frac{A^{1/2}}{x} \tag{5-2}$$

where

V_x = centerline velocity at distance x from diffuser, fpm (m/s)

V_o = average outlet velocity at diffuser, fpm (m/s)

K = velocity decay coefficient, dimensionless

A = diffuser effective outlet area, ft^2 (m^2)

x = distance to centerline velocity V_x, ft (m)

The K-value in Equation 5-2 is required to determine the separation distance in Equation 5-1. The K-value depends on the type of diffuser producing the jet. It is determined by the manufacturer from velocity profile measurements and is usually reported with the diffuser specifications. If the K-value for a specific diffuser cannot be obtained from the manufacturer, it can be estimated. Representative velocity decay coefficients for a variety of diffusers are tabulated in chapter 31, "Space Air Diffusion," of the *1993 ASHRAE Handbook—Fundamentals* (ASHRAE 1993) and are listed in Table 5-6 (Tuve 1953; Koestel et al. 1950). These values

TABLE 5-6 VELOCITY DECAY COEFFICIENT SELECTION

Diffuser Type	K-value
Circular	1.1
Slot	5.5
Jet	7.0
Perforated	3.7-4.9

can be used as an estimate for initial calculation and preliminary design purposes.

The *K*-value is related to the throw, *T*, of the jet, as shown in Equation 5-3. Equation 5-3 is a rearrangement of Equation 5-2, with a substitution of the outlet flow rate for the outlet velocity. Equation 5-3 indicates that for the same outlet flow rate, a diffuser with a lower *K*-value will have a lower throw than a diffuser with a larger *K*-value.

$$T = \frac{KQ_o}{V_x A^{1/2}} \qquad (5\text{-}3)$$

The entrainment of the jet is modeled with Equation 5-4, which is an integrated version of Equation 5-2. The equation indicates that the amount of room air entrained in the jet increases linearly with distance downstream. The entrainment is also inversely proportional to the *K*-value. This indicates that as the *K*-value decreases, the entrainment increases.

$$\frac{Q_x}{Q_o} = \frac{2x}{KA^{1/2}} \qquad (5\text{-}4)$$

where

Q_X = volume flow at distance x from diffuser, cfm (L/s)

Q_O = diffuser outlet volume flow rate, cfm (L/s)

There is a common misconception that high-entrainment jet mixing is necessary for cold air distribution. The actual situation is more complex. High-entrainment diffusers for cold air distribution will have relatively rapid temperature mixing but a relatively short throw, since their *K*-value is relatively low. Therefore, "high entrainment" implies a relatively small *K*-value and low throw values. The throw values should meet the throw recommendations given in Table 5-3. Additional information about jet entrainment is given in chapter 31, "Space Air Diffusion," of the *1993 ASHRAE Handbook—Fundamentals* (ASHRAE 1993) and in Tuve (1953).

5.7 GUIDELINES FOR COLD-AIR DIFFUSER SELECTION

5.7.1 INTRODUCTION

In this section an expanded form of the ASHRAE diffuser selection procedure is presented (ASHRAE 1993, chapter 31). The diffuser selection procedure for cold-air systems is basically the same as for conventional systems, with the additional consideration for the separation distance.

The main concept in diffuser selection is the comparison of the three parameters: the diffuser throw, the diffuser separation distance, and the room characteristic length. Ratios of these parameters are used to find optimum performance conditions. Satisfactory diffuser operation will result when recommended ratios of these parameters are used.

5.7.2 DIFFUSER SELECTION PROCEDURE

The following steps are recommended for selecting diffusers for a given application:

1. Determine the airflow rate requirements.
2. Select a candidate diffuser type and location in room.
3. Determine the room characteristic length, *L*.
4. Select the recommended throw/characteristic length (*T/L*) ratios.
5. Calculate the throw distance (*T*) range.
6. Choose the appropriate diffuser size to meet flow rate-throw requirements.
7. Compute the separation distances for maximum and minimum flow rates and compare with room characteristic length.
8. Check to see that other specifications such as noise and static pressure are also met.
9. Resize as necessary.

The steps are addressed in more detail in this section.

1. Determine the maximum and minimum airflow rate requirements.

The room airflow rate, Q, is a function of the room sensible cooling load, q, and room-supply temperature difference given by the energy equation, Equation 5-5. The maximum airflow for the room is determined from the peak load for the individual room and the design supply air temperature. Considerations regarding the selection of the design supply air temperature are discussed in section 8.6.

In actual use, due to variable occupancy patterns, the flow rate to meet the minimum load may be less than the required ventilation flow rate. The minimum airflow is determined by the minimum required ventilation airflow. The ventilation air requirements are discussed in chapter 3.

$$Q = \frac{q}{\{C_1(T_{room} - T_o)\}} \tag{5-5}$$

where

Q = room airflow rate, cfm (L/s)
q = room sensible load, Btuh (W)
C_1 = air density and specific heat product, $C_1 = 1.1$ (1.23)
T_{room} = average room temperature, °F (°C)
T_o = diffuser outlet temperature, °F (°C).

2. Select a candidate diffuser type and location in room.

The type of diffuser selected will depend on the preference of the architect and the building owner. Various types of diffusers are listed in manufacturers' catalogs and data sheets and are discussed in section 5-2. The location in the room will depend on the type of diffuser selected (as shown schematically in Figure 5-4), lighting fixture placement, and exterior window location.

3. Determine the room characteristic length, *L*.

The room characteristic length depends on the diffuser location and the distance to a wall or symmetry plane. The room characteristic length, *L*, is found from the diffuser layout pattern. The characteristic room lengths for linear and radial diffusers are defined in section 5-3.

4. Select the recommended throw/length (*T/L*) value.

The recommended *T/L* value is selected from Table 5-5 for a particular diffuser and specific room load.

5. Calculate the throw distance.

The throw distance required is found by multiplying the room characteristic length, *L*, found in step 3, by the *T/L* value determined in step 4:

$$T = \left(\frac{T}{L}\right)L \qquad (5\text{-}6)$$

6. Choose the appropriate diffuser size to meet flow rate-throw requirements.

For a chosen diffuser type, manufacturers' catalog data, such as those listed in Tables 5-1, 5-2, and 5-3, will list the flow rate-throw tables for various sizes. Given the required diffuser throw and flow rate range, a specific size of diffuser can be identified. For the linear jet and slot diffusers, the number of linear diffusers required for a space depends on the manufacturer's recommendations. The recommendations are that the sum of the lengths of the linear diffusers be 30% to 70% of the length of the wall upon which the diffusers are mounted.

7. Compute the separation distance for the flow rate range and compare with the room characteristic length.

The separation distances for a given diffuser can be calculated from Equation 5-1. The equation requires knowledge of the maximum and minimum discharge flow rates, the supply to room temperature difference, and the diffuser K-value. Compare the calculated separation distances with the room characteristic length value for the diffuser. If the separation distance at the minimum flow rate is greater than the diffuser room characteristic, then the diffuser is acceptable.

8. Check to see that other specifications such as noise and static pressure are also met.

With the higher velocities characteristic of cold-air diffusers, it is necessary to check for the pressure drop and noise level of the diffuser. A complete discussion of NC and recommended NC values for various types of applications can be found in chapter 7, "Sound and Vibration Control," of the *1993 ASHRAE Handbook—Fundamentals* (ASHRAE 1993). The diffuser pressure drop must not exceed the pressure drop chosen for diffusers in the overall HVAC system.

9. Resize as necessary.

In many cases, diffuser selection is an iterative process. It is difficult to match diffuser types and numbers with particular room loads and ventilation requirements in a single iteration. If the throw distances are too low for the flow rate range, the diffuser size will need to be decreased or the number of diffusers decreased in order to decrease the total outlet flow area. Conversely, if the throw distances are too high for the flow rate range, the diffuser size or the number of diffusers will need to be increased in order to increase the total outlet flow area. Similar considerations apply to the static pressure and NC value for the flow rates under consideration. Alternative diffuser types can also be considered in the selection process.

5.7.3 DESIGN EXAMPLE—DIFFUSER SELECTION FOR CONFERENCE ROOM

Cold-air diffusers are to be selected for the conference room example of chapter 3. The conference room is a 15 by 30 ft (6 by 9 m) space, with a cooling load of 15,354 Btu/h (4,500 W) and a design ventilation airflow rate of 450 cfm (212 L/s). The supply air temperature to the conference room is 50°F (10°C), with a 45°F (7.2°C) cold-air system, due to the fan heat gain and duct temperature rise, as shown in section 4-5. The space temperature setpoint is 75°F (23.9°C).

1. Determine the maximum and minimum airflow rate requirements.

The maximum airflow rate is determined from Equation 5-5:

$$Q = \frac{15,354}{\{(1.1)(75-50)\}} = 558 \text{ cfm} \qquad \left(Q = \frac{4,500}{\{(1.23)(23.9-10)\}} = 263 \text{ L/s}\right)$$

The maximum flow rate requirement is 558 cfm (263 L/s) to meet the cooling load. The ventilation flow rate is 450 cfm (212 L/s), so the minimum flow rate to the room is 450 cfm (212 L/s).

2. Select a candidate diffuser type and location in room.

The candidate diffuser type for this example is a ceiling slot diffuser. The slot diffuser is 48 in. (122 cm) long, with a 1-in. (2.54-cm) slot width. The slot diffusers are to be located on the edge of the room, parallel to the major axis, as shown in Figure 5-4.

3. Determine the room characteristic length, *L*.

The room characteristic length, *L*, is 15 ft (4.6 m) for the slot diffusers, as indicated in Table 5-4.

4. Select the recommended *T*/*L* ratios.

From Table 5-5, for a room load of 35 Btu/ft^2 (110 W/m^2) the acceptable *T*/*L* ratios are 0.5 to 3.3 for the slot diffuser.

5. Calculate the throw distance range.

The required throw distance range is the *T*/*L* ratio range multiplied by the room characteristic length.

From Equation 5-6:

$$T = \left(\frac{T}{L}\right)L$$

$T/L = 0.5$ to 3.3,
$L = 15$ ft. (4.6 m),
$T = 7.5$ to 49.5 ft (2.3 to 15.2 m).

6. Choose the appropriate diffuser size to meet flow rate-throw requirements.

Three 48 in. (122 cm) long slot diffusers are to be used, for an active length of 12 ft (3.7 m), about 40% of the wall length. The locations in the room are shown schematically in Figure 5-4.

From the data given in Table 5-2, the slot diffuser has a throw of about 26 ft (7.9 m) at its maximum flow and a throw of about 24 ft (7.3 m) at the minimum flow. Since the recommended throw distances for a slot diffuser in this room are from 7.5 ft to 49.5 ft (2.3 m to 15 m), the diffuser meets the throw requirements over the flow rate range.

7. Compute the separation distance for the minimum flow rate and compare with the room characteristic length.

Since the separation distance is proportional to the flow rate, the shortest separation distance will occur at the minimum flow rate. To obtain the separation distance, Equation 5-1 is used.

$C_S = 1.2$
$K = 5.5$ (from Table 5-6)
$a = 1.47$ (0.0689)
$\Delta T = 25$°F (13.9°C)
$T = 535$°R (297 K)
$Q = 150$ cfm (70 L/s)
$\Delta P = 0.10$ in. w.g. (25 Pa) (from Table 5-2)

$$x_s = aC_S K^{1/2}\left(\frac{\Delta T}{T}\right)^{-1/2} Q^{1/4} \Delta P^{3/8}$$

$$= (1.47)(1.2)(5.5)^{1/2}\left(\frac{25}{535}\right)^{-1/2}(150)^{1/4}(0.10)^{3/8} = 28 \text{ ft.}$$

$$\left(x_s = (0.0689)(1.2)(5.5)^{1/2}\left(\frac{13.9}{297}\right)^{-1/2}(70)^{1/4}(25)^{3/8} = 8.7\text{m}\right)$$

The separation distance of 28 ft (8.7 m) is greater than the characteristic length of 7.5 ft (2.3 m), so the separation distance performance is acceptable.

8. Check to see that other specifications such as static pressure and noise are also met.

From Table 5-2, the slot diffuser has a static pressure drop of about 0.14 in. (35 Pa) at its maximum flow rate of 186 cfm (80 L/s). If the system design allows a maximum diffuser static pressure drop of 0.25 in. (62 Pa), the slot diffuser is acceptable.

The NC values are to be at or below 40. From Table 5-2, the slot diffuser has an NC value of about 30 at its maximum flow rate of 186 cfm (80 L/s).

9. Resize as necessary.

For this example, no resizing of the linear slot diffusers was needed.

5.8 MIXING BOXES

Mixing boxes are used to mix room air or return plenum air with the primary supply air at a location near the room diffuser. The purpose of mixing boxes is to maintain a constant volume of air to the occupied space. As the cooling load decreases in a space, the mixing boxes will increase the amount of plenum air mixed with the primary supply air. The temperature of the air supplied to the space will rise, but the volume flow rate to the space is kept as constant as possible. If the volume flow rate to the occupied space is kept constant, the mixing of the room air by the supply air will maintain acceptable thermal comfort conditions.

Mixing boxes have been used in cold-air distribution systems to increase the final supply temperature to that of conventional systems. This approach allows the designer to use familiar procedures for diffuser selection. Mixing boxes may not be necessary in systems with diffusers selected for low airflow and cold supply air. With the proper choice of a diffuser, the possibility of jet separation is reduced, and adequate jet mixing will occur before the cold air enters the occupied space. In addition, if the diffuser selection procedure includes the minimum ventilation airflow requirements, the diffuser will provide adequate room air motion over the flow rate range.

Terminal mixing boxes are classified as induction, parallel, and series. The induction boxes operate by entrainment of plenum air. The parallel boxes have a fan operating in parallel with the main air handler, and the series boxes have a fan operating in series with the main air handler.

A serious problem with fan-powered mixing boxes is that continuously operating, small, low-efficiency mixing-box fans can use enough power to cancel out the fan power savings from a cold-air distribution system (Elleson 1993). The fan power required for a mixing box can be obtained from manufacturers' data or from ARI ratings. Many simulations of building energy performance do not accurately consider the energy consumption of fan-powered mixing boxes.

5.8.1 INDUCTION TERMINALS

The induction terminal unit is used to mix the system primary air with room air before delivery to the room. The purpose of the induction terminal unit is to maintain the volume flow rate of air to the room when the primary airflow rate is reduced. The induction terminal behaves in a manner similar to that of an in-room diffuser, which produces a low-pressure jet of air to entrain room-temperature air.

A schematic of a representative induction mixing box is shown in Figure 5-7. The induction terminal is typically located above the ceiling in the return air plenum. The modulating damper in the induction box lowers the primary airflow static pressure. The lowered static pressure induces higher pressure plenum air into the induction box. The induced air is supplied directly from the return air path through an induction damper. If the room is receiving too much cooling, the room thermostat will direct the control system to reduce the primary air volume by closing the primary modulating damper and opening the induction damper. An increased amount of room air is then induced into the terminal, added to the primary airstream, and delivered to the room diffuser.

Advantages of induction boxes are their simple operating strategy and easier diffuser selection due to approximately constant total flow to diffusers for 70% to 100% of primary airflow. When full cooling of a room is required, the ventilation system maintains the primary air volume at the induction box and closes the induction damper. The mixing in the induction box raises the air temperature and flow rate at the terminal diffuser without the added expense and complication of additional fans.

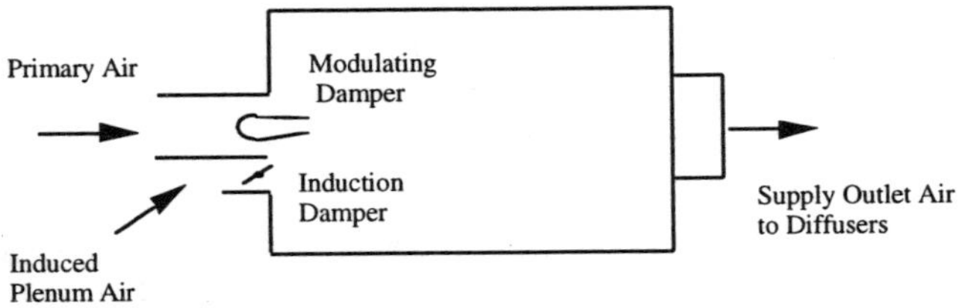

FIGURE 5-7 Schematic of Induction Mixing Box

Disadvantages of induction boxes are that they represent an additional cost and maintenance item for the HVAC system; they require higher inlet static pressure, causing increased system fan heat gain and energy use; they have increased pressure drop due to the heating coil located in the primary airflow and a minimum turndown in the range of 20% to 40% of total maximum flow. Induction boxes should not be oversized, as boxes selected too large tend not to perform well at low loads.

The amount of additional airflow due to induction in the box is shown in Figure 5-8. The figure is developed from a manufacturer's data sheet for a representative induction terminal. The data sheet typically will list the flow rate fraction of discharge air as a function of the primary airflow rate fraction for a specified inlet static pressure and downstream static pressure. The induction rate is nonlinear, since the local static pressure difference, which creates the induction, is proportional to the square of the airflow rate. For primary airflow rate fractions at or above 60%, the fraction is typically above 80%. At 50% primary air, the induction box induces an additional 21% room air to bring the delivery air to 71%; at 20%

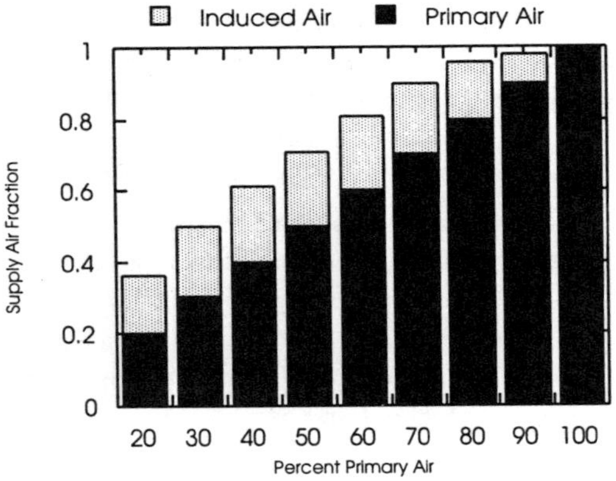

FIGURE 5-8 Induction Box Characteristics.

primary air, the induction box induces an additional 18% to bring the delivery air to 38%.

The discharge temperature from an induction terminal can be found from a sensible energy balance on the induction box, Equation 5-7. If the room air is assumed to be 75°F (24°C) and the primary air is 40°F (4.4°C), at 50% primary air, the induction box increases the delivery air to 50°F (10°C), and at 20% primary air, the induction box increases the delivery air to 56°F (13°C).

$$T_{supply} = \frac{\{(T \cdot F)_{primary} + (T \cdot F)_{induced}\}}{F_{supply}} \tag{5-7}$$

where

T = temperature, °F (°C)

F = fraction, from Figure 5-8

For example, for the case of 50% design airflow:

$$T_{supply} = \frac{\{(40)(0.50) + (75)(0.21)\}}{0.71} = 50°F$$

$$\left(T_{supply} = \frac{\{(4.4)(0.50) + (24)(0.21)\}}{0.71} = 10°C\right)$$

5.8.2 PARALLEL FAN-POWERED BOXES

Parallel fan-powered mixing boxes mix the primary and room air downstream of the fan. As shown in Figure 5-9, the fan is located outside the primary airstream between the plenum air and the box, as it is used to draw plenum air into the mixing box. Since the airflow through the fan is plenum air only, it is smaller than a series fan. Parallel fan-powered mixing boxes require a higher static inlet pressure, typically 0.5 in. w.g. (124 Pa), than series boxes since the primary air is not drawn through the mixing box. In a typical conventional applications, the fan only operates at low primary airflow rates.

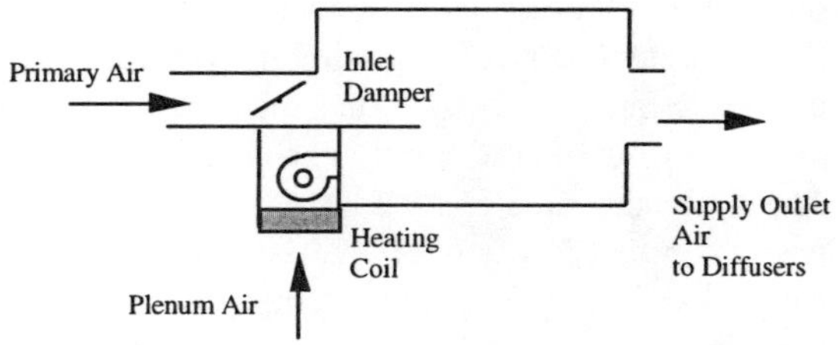

FIGURE 5-9 Schematic of Parallel Fan-Powered Mixing Box.

Parallel boxes can be useful in exterior or perimeter spaces that require heating while interior zones require cooling and in perimeter spaces where loads vary. An optional heating coil is used to heat the plenum air. As cooling load and primary airflow drop, the fan is activated, and, when necessary, heating elements can also be activated

There are two approaches to the selection of parallel fan-powered mixing boxes. One is to select the fan for 40% of the supply air to the diffusers, based on a 55°F (12.8°C) supply air temperature. The primary air would then supply 60% of the supply air to the diffusers. In this case, the fan runs continuously. This arrangement significantly reduces the fan power compared to the series box. The airflow through the diffuser is variable due to the low volume at minimum primary airflow. This approach is not recommended where heating is required.

The other approach is to select the fan for intermittent operation. This is the traditional parallel fan-powered box design where the fan is selected for heating only. After the primary cooling air has been reduced to the minimum required for ventilation and the temperature has decreased to the heating setpoint, the fan is used to provide the first stage of heat using plenum air. The design heating air temperature should be between 90°F (32.2°C) and 100°F (37.8°C).

The parallel fan box has increased complexity over the series box due to the need to turn the fan on and off at different primary airflow rates. Some care must also be used in diffuser selection with a parallel fan-powered box because of the varying flow rates through the diffuser and to meet both cooling and heating requirements.

5.8.3 SERIES FAN-POWERED BOXES

Fan-powered series mixing boxes are mounted above the ceiling with both primary and room airflow through the box fan. The fan is selected for the air quantity that would be required for 55°F (12.7°C) supply air. As shown in Figure 5-10, room air is drawn by the series fan into the box and mixed with the cold primary air upstream of the fan. Control is achieved with an inlet air damper. The series fan runs continuously, supplying the diffusers with a nearly constant flow rate of air. The inlet pressure required is typically 0.1 in. w.g. (25 Pa) to overcome losses in the primary inlet air damper. An optional heating coil is located downstream of the series fan.

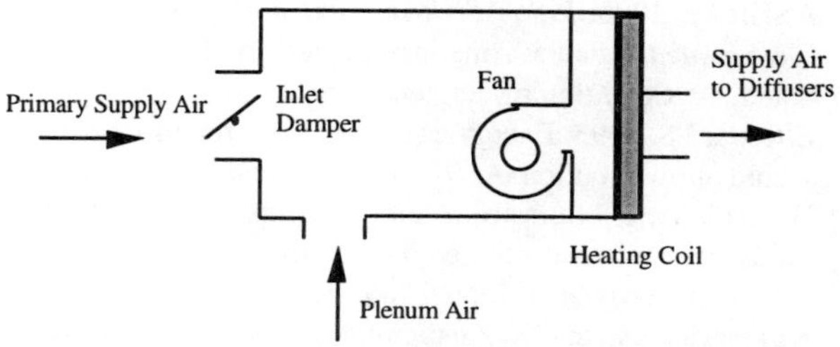

FIGURE 5-10 Schematic of Series Mixing Box.

Advantages of series mixing boxes are ease of diffuser selection, design, and control, and convenient reheat/morning warm-up capability. The sound level radiated from the box is essentially constant. They can provide heating and nearly constant flow to the diffusers over the entire operating range. The constant flow rate at low primary air temperatures increases room air motion and can alleviate possible problems with jet separation.

Disadvantages are increased initial cost, energy use, and noise. For the same zone size, series mixing boxes tend to have a higher initial cost since the fan is sized to accommodate total airflow for the zone rather than just the heating airflow. Noise levels in the occupied space are increased due to the extra equipment running above the ceiling. Series boxes, as with any extra mechanical devices, add complexity and maintenance needs to the system.

5.8.4 SELECTION OF FAN-POWERED MIXING BOXES

The selection procedure for fan-powered mixing boxes is detailed in manufacturers' catalogs. At present, there is no overall procedure for the selection of fan-powered mixing boxes published in the *ASHRAE Handbook.* The selection procedure generally involves determination of the primary and plenum air requirements for a given supply outlet flow rate and temperature. Appropriately sized primary air damper and series or parallel fans are then chosen to meet the flow rate-static pressure requirements.

5.9 REFERENCES

ASHRAE. 1993. *1993 ASHRAE handbook—Fundamentals.* Atlanta: American Society of Heating, Refrigerating and Air-Conditioning Engineers, Inc.

ASHRAE. 1996. *1996 ASHRAE handbook—HVAC systems and equipment*. Atlanta: American Society of Heating, Refrigerating and Air-Conditioning Engineers, Inc.

Elleson, J.S. 1993. Energy use of fan powered mixing boxes with cold air distribution. *ASHRAE Transactions* 99(1): 1349-1358.

Hassani, V., T. Malmstrom, and A. Kirkpatrick. 1993. The indoor thermal environment of cold air distribution systems. *ASHRAE Transactions* 99(1): 1359-1365.

Kirkpatrick, A., and K. Knappmiller. 1996. The ADPI of cold air jets. *ASHRAE Transactions* 102(1).

Knappmiller, K., and A. Kirkpatrick. 1994. Computational determination of the behavior of a cold air ceiling jet in a room with a plume. *ASHRAE Transactions* 100(2): 677-684.

Koestel, A., P. Hermann, and G. Tuve. 1950. Comparative study of ventilating jets from various types of outlets. *ASHVE Transactions* 56: 459-478.

Li, Z., L. Christianson, J. Zhang, and A. Zhivov. n.d. Testing commercial air diffusers with cold supply air. *ASHRAE Transactions* (in review).

Miller, P. 1991. Diffuser selection for cold air distribution. *ASHRAE Journal* 33(9): 32-36.

Nielsen, P. 1987. Measurements of buoyant wall jet flows in air conditioned rooms. *Room Vent 87*, Stockholm.

Rodhal, E. 1977. The point of separation for cold jets flowing along the ceiling. *Clima 2000*, Belgrade, pp. 219-228.

Tuve, G.L. 1953. Air velocities in ventilating jets. *ASHVE 'Transactions* 59: 261-282.

Zhang, J.S., R. Zhang, Z. Li, C.Y. Shaw, L.L. Christianson, and L.H. Sparks. 1994. An experimental study of the ventilation performance of cold-air distribution systems. *ASHRAE Transactions* 100(2): 360-367.

CHAPTER SIX

CONTROL STRATEGIES

6.1 INTRODUCTION

The primary purposes of an air-conditioning system are to supply adequate ventilation air to, and to control the dry-bulb temperature in, each of the zones served by the system. A secondary purpose is to control relative humidity. These goals are typically accomplished by controlling the following parameters:

- volume of outside air supplied by the central system,
- total volume of air supplied by the central system,
- temperature of air supplied by the central system,
- volume of air supplied to each zone, and
- temperature of air supplied to each zone.

In a typical VAV system, a zone volume control terminal modulates the volume of conditioned air supplied to the zone, in order to satisfy the zone temperature sensor. A variable-speed supply fan, discharge damper, or fan inlet vane modulates the total volume of air supplied by the system to maintain a duct static pressure setpoint. A cooling coil valve modulates the volume of fluid supplied to the coil to maintain the system supply air temperature at its setpoint. The fluid temperature is determined by a variety of chiller and central plant control sequences. An outside air damper, controlled by pressure relationships, a fan-tracking schedule, or air volume measurements, modulates to maintain the required volume of outside ventilation air. The various temperature and pressure setpoints may be modified according to appropriate reset schemes to optimize system operation. There are many variations and modifications to the standard VAV control systems that are beyond the scope of this guide. Additional coverage of the various control systems is contained in chapter 42 of the *1995*

ASHRAE Handbook—HVAC Applications (ASHRAE 1995) and in Hittle and Haines (1993).

The primary issues pertaining to the control of cold-air distribution systems are related to differences from conventional systems, including:

- control system requirements,
- control with heating loads or low cooling loads,
- preventing condensation when starting up after an overnight or weekend shutdown, and
- resetting supply air temperature to
  - minimize total energy consumption,
  - minimize reheat,
  - increase air motion at low loads,
  - maximize economizer cooling.

6.2 CONTROL REQUIREMENTS

The control requirements for basic operation of a cold-air distribution system are generally the same as the requirements for a conventional system. Cold-air distribution offers the opportunity to reduce overall energy use by optimizing the supply air temperature, but some increased sophistication in controls is required.

A direct digital control (DDC) system allows the use of more sophisticated control logic than typically used in pneumatic control systems. DDC can be applied to both the zone controls and the control of the cooling plant and air-handling unit. Initial start-up control sequences, reset schedules, and setpoints often require continuous modification to optimize system performance. DDC controllers are able to control to different parameters for different times of day. Without sophisticated control logic, the ability to determine the optimum level of reset at any time is reduced.

6.3 MINIMUM VENTILATION AIRFLOW AND REHEAT

The minimum ventilation flow rate is determined by the occupancy of the zones and is discussed in chapter 3. Reheat may be required when ventilation air requirements of the space are higher than the cooling airflow requirements of the space during low cooling loads. Reheat of the supply air is accomplished by using a heating coil in the duct or in the terminal box to raise the supply air temperature. Since reheat of the supply air has an energy penalty, the supply air setpoint should be reset to a temperature that will minimize or eliminate the need for reheat.

6.4 SUPPLY TEMPERATURE RESET

Supply temperature reset is a strategy used to maximize the cold-air distribution system efficiency. The optimum supply temperature depends on performance curves of the chiller, pump, and supply fan. As the building load changes, the optimum supply air temperature to minimize energy consumption will also change. At a specific building load, increased supply air temperature will increase the air-side energy use, decrease the dehumidification and chiller energy, and increase the economizer availability. The optimum supply air temperature for a specific load can be predicted using the modeling and simulation considerations discussed in chapter 9. Additional information regarding the dependence of the cold-air system energy use on the supply air temperature is given in Dorgan and Dorgan (1995).

A supply air reset sequence will reset the supply air temperature to a temperature below a maximum that will satisfy the zone with the highest cooling load. One supply temperature reset strategy is directed at minimizing the need for reheat during low-load conditions. As the load is reduced, the airflow rate is reduced until a minimum airflow rate is reached. If the load is reduced further, the flow rate is maintained at the minimum airflow rate, and the supply air temperature, T_{SA}, is increased.

Supply temperature reset strategies can also be used to maximize economizer availability (discussed further in section 6-7), depending on the cooling load and outdoor air temperature. If the chiller energy use is the dominant component at a given load condition, the supply air temperature should be reset upward to minimize the chiller energy use. In this case, one possible reset strategy is to operate at 55°F (12.7°C) and then ramp down to 45°F (7.2°C) as the cooling load increases. In humid climates, a maximum temperature reset increase will be desirable. This will retain the cooling system dehumidification capability under wet, low cooling load conditions, such as a rainy spring day.

Supply air reset is also used in situations that require a constant airflow to diffusers at low loads. As the supply airflow rate decreases, the room air circulation provided by the cold air diffusers will decrease. The room air circulation required is specified by the Air Diffusion Performance Index (ADPI), discussed in section 5-4. As noted in that section, diffusers that have satisfactory performance over a wide range of supply airflow rates are available.

The supply temperature reset is accomplished by control of the cooling coil temperature, which can be done using a proportional control sequence. More sophisticated types of control routines for supply temperature reset can be programmed into a DDC controller.

6.5 HEATING

Morning warm-up and night heating may be required in some climates where there is not enough heat load at night to maintain an acceptable building or perimeter zone temperature. Heating may also be required when there is not enough heat gain during the day, which also occurs in exterior zones and in small buildings. Heating can be accomplished with baseboard heaters; with heating coils located in the air ducts, VAV boxes, or fan-powered boxes; or by heating the primary air supply and modulating zone air volume for zone control. Baseboard heaters are an acceptable choice if ventilation air is preheated at the air-handling unit. Baseboard heaters would be less advantageous if zone reheat coils are needed to heat the ventilation air.

The control considerations for the heating components of a cold-air distribution system are basically the same as for a conventional system. In the heating mode, the supply air temperature should be reset to the maximum level, after the terminal unit's primary airflow is set at the minimum position, if baseboard radiation or reheat is used. The performance of the cold air diffusers in a heating mode must be reviewed, as discussed in section 5-4. The difference between the heating supply air temperature and the room temperature should be kept less than 20°F (11 °C).

6.6 RECOVERY FROM NIGHT/WEEKEND SETBACK

During unoccupied hours, the cooling systems in office buildings are customarily shut down. The resulting interior space temperature and humidity during this time will depend on the interior load and the outside environmental conditions. In climates where the outside humidity is greater than the controlled interior humidity, there will be an increase in the interior humidity levels. The mass transfer processes include both infiltration and direct vapor transfer through the building envelope. Both of these transport processes should be relatively slow whenever proper vapor or moisture barriers are installed and the building envelope is tight, including windows and doors.

The increased interior humidity can result in condensation problems at start-up. One method to mitigate possible condensation problems is to initially operate the cooling system at higher temperatures. Spaces subject to large amounts of infiltration, such as entry and lobby areas, can be served by fan-powered boxes or by separate air-handling units with higher supply temperatures. Optimal airflow rates and starting times will vary for each system and depend on the fan power use characteristics, chiller or cooling system energy use, and occupancy schedules.

The reset of the supply air temperature from the setback to the normal setpoint can be accomplished using either a temperature ramp-down schedule or a measurement of the return air dew point. A representative example of a temperature ramp-down schedule is shown in Table 6-1.

TABLE 6-1 RAMP-DOWN EXAMPLE

Time	Airflow Limit	Delivery Temperature
Two hours before occupancy	40% of Maximum	55°F (13°C)
One hour before occupancy	65% of Maximum	50°F (10°C)
Occupied	100% of Maximum	45°F (7°C)

6.7 CONTROL OF CONDENSATION

Condensation will occur on a diffuser surface if its surface temperature is at or below the dew-point temperature. Cold-air systems have a lower dew-point temperature than conventional systems due to the additional moisture removed at the coiling coil. As shown in Figure 3-3, the dew-point temperature is about 49°F (9.5°C) for a space with a dry-bulb temperature of 75°F (24°C) and a relative humidity of 40%. By way of comparison, this is 6°F (3°C) below the dew-point temperature of conventional systems.

There are two diffuser surfaces of concern: the room surface and the plenum surface. The room-side diffuser surface will be at some intermediate temperature between the supply temperature and the room temperature. As discussed in Dorgan and Elleson (1988), the room surface temperature of metal diffusers is generally 3°F (2°C) warmer than the supply air. For a plastic diffuser, the temperature difference is up to 10°F (6°C). The minimum diffuser supply air temperature for condensation control can be found from the following equation:

$$T_o = T_{dp} - dT_{fan} - dT_{duct} - dT_s \tag{6-1}$$

where

T_o = diffuser outlet temperature, °F (°C)

T_{dp} = space dew-point temperature, °F (°C)

dT_{fan} = fan temperature rise, °F (°C)

dT_{duct} = duct temperature rise, °F (°C)

dT_s = diffuser surface and supply air temperature difference, °F (°C)

For example, for T_{dp} = 49°F (9.5°C), dT_{fan} = 2°F (1°C), dT_{duct} = 3°F (1.5°C), and dT_s = 3°F (2°C), the minimum diffuser outlet temperature, T_o, is 41°F (5°C). Example 4.5.1 indicates that the diffuser outlet temperature for a 45°F (7°C) cold-air system is 50°F (10°C). This diffuser outlet temperature is above the minimum indicated by Equation 6-1. Therefore, the condensation potential on the diffuser surface for these conditions is minimal.

On the plenum side of the diffuser, the ambient humidity may be higher than on the room side, especially in systems with ducted returns additional insulation is required. The diffuser insulation requirements to prevent condensation on the plenum side are the same as those for the ductwork. These considerations are discussed in section 4-6.

6.8 ECONOMIZER-CYCLE CONTROL

Economizer-cycle operation involves bringing outdoor air to the cooling coil instead of return air when the outdoor air has a lower enthalpy than the return air. This will result in decreased mechanical cooling energy requirements. The outdoor air temperature, which has the same enthalpy as the return air, is termed the "switch-over" temperature.

The availability of the economizer cycle depends on outside environmental conditions. Cold-air systems have reduced economizer availability due to their lower supply air temperature and humidity. For a cold-air system, less cooling can be obtained from outside air when the outside temperature is between the conventional supply air temperature and the cold-air supply temperature. For example, when the outside air temperature is 50°F (10°C), the conventional system can meet the entire load from the outside air, but the cold-air system must provide some mechanical cooling to meet its supply air setpoint. Some options to reduce this problem are to reset the supply air temperature based on load or outside air temperature or to override the supply setpoint during economizer operation.

The switch-over temperature is usually lower in a cold-air system than a conventional system because the return air in a cold-air system has a lower humidity and therefore a lower enthalpy. As a consequence, there will be some hours during operation when some cooling would be available for a conventional system, but not a cold-air system. If the supply air is reset to 55°F (13°C), the return air condition will be the same as that of a conventional system.

Enthalpy sensors can be used to measure the enthalpy of the outdoor and return air, allowing the control system to select the appropriate mode. However, enthalpy sensors are expensive, require calibration, and need to be calibrated every three to six months. The need for an enthalpy calculation from dry-bulb and relative humidity adds additional complexity to the control system.

A dry-bulb switch-over control can operate more reliably, and nearly as effectively, as an enthalpy control. This control sequence activates the economizer cycle when the outdoor dry-bulb temperature falls below the switch-over temperature. The dry-bulb switch-over temperature is selected to correspond on average to the condition at which the outdoor enthalpy is below the return air enthalpy. Dorgan and Dorgan (1995) report that an optimum dry-bulb switch-over temperature may be as high as 78°F (26°C) in a low-humidity environment such as Arizona and as low as 62°F (17°C) in a more humid environment such as Washington, D.C.

6.8.1 DESIGN EXAMPLE—DETERMINATION OF SWITCH-OVER TEMPERATURE

The following example illustrates the selection of the switch-over temperature for 45°F (7°C) and 55°F (13°C) supply air systems, for Madison, Wisconsin weather data.

1. Determine the return air condition.

For this example, assume the cold air return wet-bulb temperature is 60°F (15.5°C) and the conventional system wet-bulb temperature is 62.8°F (17.1°C). These wet-bulb temperatures result from the psychrometric analysis in section 4-10. The specific conditions are as follows:

Cold-air system: 76.5°F (24.7°C) db, 40% RH, 26.5 Btu/lb$_a$ (61.6 kJ/kg$_a$) enthalpy, 60.2°F (15.6°C) wb.

Conventional system: 75.5°F (24.1°C) db, 50% RH, 26.5 Btu/Ib$_a$ (65.8 kJ/Kg$_a$) enthalpy, 62.8°F (17.1°C) wb.

2. Determine the mean coincident wet-bulb temperatures associated with a range of outdoor dry-bulb temperatures.

Table 6-2 is a tabulation of weather data for Madison, Wisconsin, showing the mean coincident wet-bulb temperature for each 5°F (3°C) range of dry-bulb temperatures. For example, in June the average wet-bulb temperature is 75°F (24°C) when the dry-bulb temperature is between 90°F (32°C) and 94°F (34°C).

The cold-air system should use economizer cooling when the outdoor wet-bulb is below the return wet-bulb temperature of 60°F (15.5°C). The mean coincident wet-bulb is at or below 60°F (16°C) for the 60°F to 64°F dry-bulb bin for most months. A dry-bulb switch-over setting of 65°F (18°C) will correctly provide economizer cooling for most conditions.

The conventional system should change over to economizer cooling when the outdoor wet-bulb temperature is below the return wet-bulb temperature of 63°F (17°C). The mean coincident wet-bulb is 64°F (18°C) or below in each month for the 65°F to 69°F dry-bulb bin. The switch-over temperature for the conventional system is estimated as 67°F (19°C).

Note that if the dry-bulb switch-over were set at 78°F (25°C), based on a simple comparison between the outdoor and return air dry-bulb temperatures, the system would be using outdoor air with a higher enthalpy than the return air for a large number of hours. For this location, the changeover temperature could be set lower in

TABLE 6-2 REPRESENTATIVE MEAN COINCIDENT WET-BULB (I-P UNITS)

Dry-Bulb (°F)		Mean Coincident Wet-Bulb (°F)						
Bin		April	May	June	July	August	Sept.	October
90	94			75		80		
85	89			71	74	79	73	
80	84		68	68	70	73	70	
75	79		64	66	68	70	67	64
70	74	55	61	64	67	68	65	62
65	69	54	58	62	64	64	62	58
60	64	48	55	58	60	61	58	56
55	59	46	52	54	56	56	54	52
50	54	44	48	49	51	52	50	49
45	49	41	44	45	47	47	46	43
40	44	38	40	42		43	42	40
35	39	35	36				38	34
30	34	31	30				33	29
25	29	27	27					24
20	24	19						19

(SI UNITS)

Dry-Bulb (°C)		Mean Coincident Wet-Bulb (°C)						
Bin		April	May	June	July	August	Sept.	October
32	34			24		27		
29	32			22	23	26	23	
27	29		20	20	21	23	21	
24	26		18	19	20	21	19	18
21	23	13	16	18	19	20	18	17
18	21	12	14	17	18	18	17	14
15	18	9	13	14	16	16	14	13
13	15	8	11	12	13	13	12	11
10	12	7	9	9	11	11	10	9
7	9	5	7	7	8	8	8	6
4	7	3	4	6		6	6	4
2	4	2	2				3	1
-1	1	−1	−1				1	−2
-4	−2	−3	−3					−4
-7	−4	−7						−7

April and May if the control system has the capability of resetting the setpoint according to the time of year.

6.8.2 DESIGN EXAMPLE—ECONOMIZER AVAILABILITY DETERMINATION

The availability of the economizer cycle depends on outdoor air temperature profiles and the switch-over temperature. The switch-over temperature for a given system is determined in section 6.8 The hours per year when the outdoor air dry-bulb temperature is below a switch-over temperature can be found from a cumulative hour curve for the outdoor air temperature profile. The outdoor air temperature dry-bulb temperature profile can be obtained from the bin weather data for a particular site.

Figure 6-1 is a representative graph of cumulative bin weather data for a given location. The cumulative curve is constructed by consecutively adding the bin temperature data from low to higher temperatures. Figure 6-1 indicates that at that particular location there are 6,000 hours below a temperature of 60°F (16°C). For the cold-air system with a recommended switch-over temperature of 65°F (18°C), there are 6,750 hours below 65°F (18°C). For the conventional supply air temperature system with a recommended switch-over temperature of 67°F (19°C), there are 7,050 hours below 67°F (19°C). The cold-air system has a reduced economizer availability of about 300 hours.

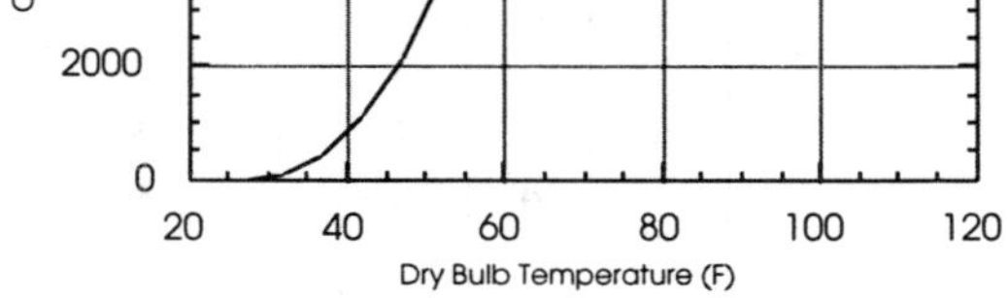

FIGURE 6-1 Sample bin weather data.

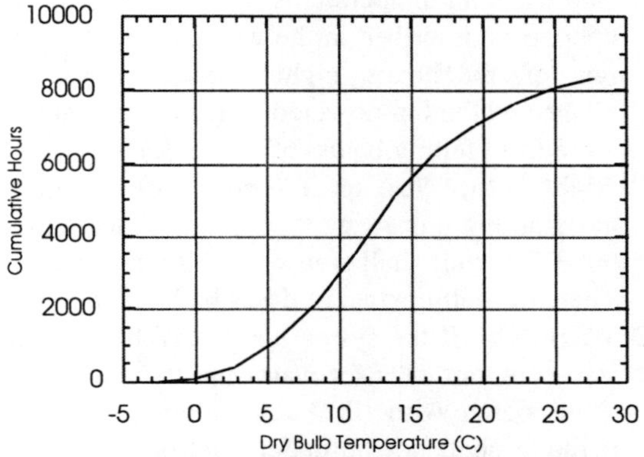

FIGURE 6-1 (SI) Sample bin weather data.

6.9 CONTROL SEQUENCES

6.9.1 DESCRIPTION OF CONTROL SEQUENCES

Control sequences are graphical or written representations of how an HVAC system is intended to react to given inputs. This information is used by vendors to supply bids and build the control system plus programming for the system. Other control sequence documentation users will be system operators and the system designer to ensure that all required equipment for the intended operation is included in the design. A system may have more than one control sequence, such as a night setback control sequence, a soft-start sequence, and a standard operating sequence.

6.9.2 EXAMPLE CONTROL SEQUENCE

This example shows an outline for specification of a simple control sequence for a cold-air distribution system.

A. **Unoccupied**: When the building is indexed for unoccupied operation, the unit supply fan shall be stopped, the chilled-water valve position shall be closed, the outside air damper shall be positioned closed, the return air damper shall be positioned open, and the smoke damper shall remain open. All associated exhaust fans shall be stopped.

B. **Night Setback**: If the system uses VAV boxes, the space temperature sensor shall signal the air-handling unit to start when any space temperature drops to 60°F (15°C). The unit shall stop when all temperatures rise to 65°F (18°C). The unit shall operate as described in the warm-up mode. If the system uses fan-powered boxes, night setback control shall be accomplished by the fan-powered boxes to prevent the space temperature from falling below 65°F (18°C).

C. **Night Setup**: The space temperature sensor shall signal the air-handling unit to start when any space temperature rises to 90°F. The unit shall stop when all temperatures drop to 85°F. The unit shall operate as described under cool-down mode.

D. **Warm-Up**: If the system uses VAV boxes, when the optimal start program calls for warm-up, the unit shall be started and shall operate with 100% recirculation air. The discharge temperature shall not be under control. If the system uses fan-powered boxes, when the optimal start program calls for warm-up, the unit shall remain off and all warm-up will be accomplished by the fan-powered boxes.

E. **Cooldown**: When the optimal start program calls for cooldown operation, the unit shall be started and shall operate with 100% return air, except the economizer cycle shall function as described below. The unit shall control the cooling coil as described under "Temperature Control Reset."

F. **Occupied**: When the building is indexed for occupied operation, if the unit is not running, the unit shall be started and the unit minimum outdoor air damper, the return air damper, and the relief air damper shall be positioned to the minimum outside air position. All associated toilet exhaust and miscellaneous fans shall be started. The actual time for occupied operation shall be one hour prior to the normal occupancy time to permit an indoor air quality (IAQ) pre-operation period.

G. **Temperature Control Reset**: Upon a rise in discharge temperature setpoint, the chilled-water valve shall be modulated open. The discharge air temperature shall be initially set to control at 56°F (16°C) and ramped down to the design temperature of 43°F (6°C), based on room demand. Upon a drop in space temperature, the discharge temperature shall be reset upward in a similar manner. The setpoint shall not be changed more than 1°F (0.5°C) over any five-minute period. For example, when 10% of the terminals are at 76°F (24°C), the cooling setpoint, the discharge temperature shall be at 45°F (7°C). When all of the terminals drop below 76°F (24°C), the temperature shall start to ramp up. When 90% of the sensors are at 73°F (23°C), the temperature shall be at 56°F (16°C).

If the chilled-water valve is positioned to fully open and the discharge temperature is at the setpoint of 56°F (16°C), a signal shall be sent to the building control system to start the chillers to supplement the capacity of the water-side economizer cycle. The secondary chilled-water discharge temperature shall be set downward as required to satisfy the worst case air-handling units.

H. A high-limit economizer changeover shall return outside and relief dampers to their minimum position whenever the outside air dry-bulb temperature is above the changeover temperature, T_C. (Note: The appropriate T_C depends on climate. The appropriate changeover temperature must be stated in the specification.)

I. The initial soft-start sequence for this building shall be as listed in Table 6-3:

TABLE 6-3 SOFT-START SEQUENCE

Time	VAV Fan Setpoint	VAV Duct Setpoint	Desired Fan cfm
6:30-7:30 a.m.	0.9 in. w.g. (224 Pa)	0.23 in. w.g. (57 Pa)	40%
7:30-8:30 a.m.	2.1 in. w.g. (523 Pa)	0.53 in. w.g. (132 Pa)	65%
All other occupied periods	4.4 in. w.g. (1,100 Pa)	1.10 in. w.g. (274 Pa)	100%

6.10 MONITORING

Monitoring of system performance allows operators to judge the effectiveness of the system in maintaining comfort, to track energy consumption, and to evaluate the effects of control changes. Monitoring is particularly cost-effective for large systems with full-time HVAC operators, where small percentage changes in energy consumption can amount to large dollar savings. However, some level of monitoring, combined with automatic generation of log reports, will be beneficial even in smaller systems.

Monitoring system performance has benefits for conventional as well as cold-air distribution systems. Cold-air systems may offer more benefits since they allow more flexibility in optimizing the supply air temperature. Monitoring of zone air temperatures provides operators with an indication of how well the system is meeting space cooling load, and is a minimum monitoring requirement.

Monitoring of supply airflow to each zone is useful for balancing and troubleshooting and also allows for reset sequences based on the cooling load in individual zones.

To evaluate energy efficiency, control systems should measure and record system cooling load and energy consumption of fans, pumps, and chillers. Calculating energy efficiency in terms of kW/ton for each component and for the entire system allows operators to devise control sequences to maximize energy efficiency.

Parameters to be monitored by a building management system involve three categories: space conditions, equipment conditions, and equipment energy use. Monitoring space conditions may involve any or all of the following variables: space temperature, dew point, and ventilation rates. Equipment parameters involve properties such as cooling coil temperature and flow rate, fan speed, damper position, chiller setting, remaining ice storage capacity, and pressure drop across equipment. With complete system information it is easier to detect operating problems and inefficiencies in a system. For instance, knowledge of the fan energy and chiller energy use characteristics can help to optimize the system.

6.11 OPERATION AND MAINTENANCE

6.11.1 CONTROL SENSORS

Control sensors require periodic maintenance to function properly. Manufacturers' data provided with the equipment will give service requirements and timing for each particular sensor. Lack of maintenance on sensors can seriously degrade system performance and indoor air quality. It is important to explicitly specify maintenance and calibration requirements in the operating instructions for a building. Making building operators and managers aware of what is required will help keep a well-designed and built air-conditioning system operating up to its capabilities.

6.11.2 CONTROL SYSTEM ADJUSTMENT/OPTIMIZATION

By comparing actual energy use to expected energy use for the components of an air-conditioning system, it is possible to optimize the supply air temperature and flow rates based on actual system performance data. The control sequences should be adjusted for optimal performance of the system components.

Initial system adjustment and optimization is often specified as a requirement of the contractor providing the DDC system.

6.12 REFERENCES

ASHRAE. 1995. *1995 ASHRAE handbook—HVAC applications*. Atlanta: American Society of Heating, Refrigerating and Air-Conditioning Engineers, Inc.

Dorgan, C., and C. Dorgan. 1995. *Cold air distribution design guide*. 2nd ed. EPRI TR-105604. Palo Alto, Calif.: Electric Power Research Institute.

Dorgan, C., and J. Elleson. 1988. *Cold air distribution design guide*.1st ed. EPRI EM-5730. Palo Alto, Calif.: Electric Power Research Institute.

Hittle, D.C., and R.W. Haines. 1993. *Control systems for heating, ventilating, and air conditioning*, 5th ed. New York: Van Nostrand Reinhold.

CHAPTER SEVEN

COMMISSIONING

7.1 INTRODUCTION

Commissioning is the process by which the facility owner and design team ensure that an HVAC system performs according to its design intent. The term "commissioning" is sometimes used in reference to the start-up and performance-testing phases of a project. However, a comprehensive commissioning process spans the entire project, from concept development in the program phase through operation of the system after it is accepted.

This comprehensive approach to commissioning is described in *ASHRAE Guideline 1-1996, The HVAC Commissioning Process* (ASHRAE 1996) which provides additional detail on specific requirements for each step of the procedure and on the roles and responsibilities of the parties to the process.

Any HVAC system will benefit from some level of commissioning. For complex systems, a thorough commissioning process is essential to ensure that all subsystems and components function together in the manner intended by the design.

Commissioning is particularly important for cold-air distribution systems, which differ in many ways from conventional designs. Most designers, contractors, and operators are not familiar with the particular design parameters, installation requirements, control sequences, and operating requirements of cold-air distribution systems. Without this knowledge, these project team members are less likely to detect errors and potential problems throughout the design and installation process, to the detriment of the efficiency and effectiveness of system operation. The commissioning process provides a systematic method for the project team to avoid these problems.

The commissioning process, as described in the ASHRAE commissioning guideline, consists of the following phases:

- Program
- Design
- Construction
- Acceptance
- Training
- Operation

Readers are encouraged to consult the ASHRAE commissioning guideline for a comprehensive discussion of the commissioning process for building HVAC systems. The following sections discuss specific considerations for commissioning of cold-air distribution systems.

7.2 PROGRAM PHASE

In the program phase, the framework is developed to guide the commissioning process through the remaining phases. In this phase, the functional needs of the HVAC system are defined, design concepts are developed, and commissioning requirements and responsibilities are defined.

The two end-products of the program phase are

- a detailed statement of design intent describing the selected cooling system and
- a commissioning plan or implementation plan outlining the process by which the system will be designed, installed, and commissioned.

7.2.1 STATEMENT OF DESIGN INTENT

The statement of design intent defines the cooling system's performance requirements in detail. It is the master reference document that governs the development of the project through all of the succeeding phases.

The process of defining the design intent begins in the program phase and continues throughout the design phase. The initial statement of design intent describes the facility's functional needs, environmental control requirements, and performance requirements. Some of these items are temperature and humidity control requirements, cooling load parameters, financial criteria, and energy performance criteria. Some owner requirements that would favor the application of cold-air distribution include the need for low space humidity or a limitation on space available for ducts or fans.

A feasibility evaluation is typically carried out during the program phase to compare first costs and operating costs for a number of possible system options. Economic evaluation considerations for cold-air distribution systems are discussed in chapter 8.

The initial feasibility evaluation is often performed based on limited, tentative information about cooling loads, equipment capacities, and operating strategies. If the initial evaluation indicates that cold-air distribution is an attractive option, a more detailed study is recommended to establish accurate estimates of construction costs and operating costs.

After selecting the preferred cooling system option, the statement of design intent is expanded and refined to include specific details of system design parameters and operation.

7.2.2 COMMISSIONING PLAN

The commissioning plan describes the process that will be used to implement the project during each phase. The commissioning plan is created during the program phase and becomes more detailed during the following phases as the project takes form.

The program phase commissioning plan identifies the participants in the process and describes their responsibilities. It establishes the level of detail to which commissioning will be carried, outlining which of the items discussed in the following sections will be included. At this stage, the commissioning plan for a cold-air distribution system should describe the requirements for performance testing, including what tests are to be performed and which parties will be responsible for testing. In particular, the responsibilities of the design professional must be clearly defined and included in the designer's scope of work and fees. The design review process should also be defined.

7.3 DESIGN PHASE

Commissioning activities occurring during the design phase include a design review, continued development of the statement of design intent, and development of a commissioning specification for inclusion in design documents.

7.3.1 DESIGN REVIEW

Successful cold-air distribution projects include participation on the design team by an engineer with experience designing cold-air distribution systems. Internal or external design review is important for any cooling system of any complexity. Cold-air distribution design involves a number of considerations that do not enter into conventional design. Most designers are not experienced with these issues, and a review by an experienced engineer is invaluable.

The design reviewer should be chosen for his or her expertise and experience with cold-air distribution projects. The design reviewer should have experience not only in design but in construction and in operation and maintenance of cold-air distribution systems. The design reviewer should have experienced both success and failure in the application of cold-air distribution and must fully understand the ramifications of errors in cold-air distribution design and implementation.

Owners selecting a design reviewer should seek an individual who is a registered professional engineer and who has been involved in the design, construction management, and start-up and testing of cold-air distribution systems. The design reviewer should be knowledgeable in control of cold-air distribution systems and should have participated in previous cold-air distribution design reviews.

The design review process requires that design assumptions and calculations be well documented. This in itself tends to improve the quality of the design, since many errors or ambiguities are discovered in the course of assembling documentation for the design reviewer.

The design review may include a review of subsystems other than the cold-air distribution system. The following information pertaining to cold-air distribution should be provided to the design reviewer:

- Space load calculation including sufficient detail to understand each of the factors that affect the load
- Psychrometric analysis accounting for all system effects including
  - space latent and sensible loads
  - infiltration
  - outdoor air
  - plenum loads
  - fan heat gains
  - heat gains to ducts
  - duct leakage

- Supply fluid temperature analysis, including
  - temperature available from the cooling plant (for ice storage, the available temperature must be based on an accurate hourly design-day analysis)
  - pump and piping heat gains
  - heat exchanger performance (if applicable)
- Cooling coil performance data, including part-load performance
- Detailed drawings, including schematic diagrams, piping and instrumentation diagrams, control diagrams, plan drawings, equipment schedules, and construction details
- Control sequence description, ladder logic diagram, control specification, EMS interface specifications
- Equipment specifications—storage, chiller, pumps, cooling tower, heat transfer fluid, etc.
- Commissioning plan—start-up, acceptance testing, operator training plan and documentation, and maintenance training plan and documentation

Normally, the design reviewer is answerable directly to the owner but should be a member of the design engineer's team. It is important to maintain an open line of communication between the design reviewer and the design team to establish and foster a cooperative spirit. The key to an effective review process is to provide input to the design team when it is needed. If design review input comes too late in the design process, it can require substantial rework and possible financial loss to the designer.

7.3.2 STATEMENT OF DESIGN INTENT

During the design phase, the statement of design intent that was developed in the program phase is expanded. The following items are added to the previously defined criteria:

- Narrative description of the system
- Energy and demand performance goals
- Schematic diagram
- Control sequence descriptions

Each of these items begins in the initial design phase as a general description of the intended final result. As the design develops, they are refined and expanded so that the final statement of design intent provides a detailed description of the intended configuration, operation, and control of the system.

Specific items to be addressed in the design-phase statement of design intent for a cold-air distribution system include:

- Supply air temperature leaving the cooling coil and entering the space
- Method of delivering supply air to the space: fan-powered mixing boxes, induction boxes, or direct supply to diffusers
- Space temperature and relative humidity
- Supply fluid temperature control strategy
- Supply air temperature reset strategy
- Minimum outside air control
- Control sequences for part-load operation
- Requirements and limitations for future modifications, such as for future tenant improvement work

The statement of design intent is typically included in the construction documents to provide guidance to installation and start-up personnel.

7.3.3 COMMISSIONING SPECIFICATION

The commissioning specification defines the commissioning activities that will take place during the construction and acceptance phases of a project. The commissioning specification, which is a part of the contract documents, defines the responsibilities of each party to the commissioning process. It specifies the requirements for start-up and acceptance testing and defines the criteria by which the performance of the system will be judged.

The commissioning specification must describe commissioning requirements in sufficient detail for contractors to submit bids covering the required services. It should include the following items:

- Description of test methods
- Documentation requirements, including control system submittals, equipment submittals, test results, operation manuals, and maintenance manuals
- Relationship between completion of commissioning tasks and release of payments to contractors
- Performance criteria governing acceptance of the system
- Recourse if the required performance is not obtained
- Training requirements
- Responsibility for developing test schedules
- Responsibility for defining additional details not covered in the specification

7.4 CONSTRUCTION PHASE

Successful cold-air distribution projects benefit from teamwork and open communication among owner, designer, and contractor during the construction phase. The parties recognize that it is in the best interest of all to achieve a successful system and that cooperation and teamwork are essential to achieve that common goal.

In a typical design-bid-construct project, the construction phase begins when the successful bidder executes a contract with the owner. For the purposes of this discussion, the construction phase is considered to end when equipment has been started up, and testing, adjusting, and balancing (TAB) work has been completed. Construction-phase commissioning activities include review of submittals, progress meetings, observation of construction, and start-up.

Early in the construction phase, the contractor submits documentation on material and equipment to be installed and shop drawings illustrating details of the installation. The design engineer should carefully review this material to ensure that the installation will conform to the design intent. The review must verify that all equipment is as specified in the design documents or that any substitutions will provide acceptable, equivalent performance. Cooling coils, diffusers, VAV terminals, and duct insulation should receive particular scrutiny in the submittal review.

The primary communication channel during the construction phase is a series of progress meetings that include all members of the project team. The initial kickoff meeting is an opportunity for the parties to become acquainted and to review the requirements of the project. The design intent, documentation requirements, scheduling, and commissioning responsibilities are reviewed and clarified if necessary. While these items are also included in the design documents, it is important to verify that all parties thoroughly understand all requirements.

Observations of the progress of construction serve to verify that construction is proceeding in conformance with the contract documents, with the design intent, with equipment manufacturers' installation requirements, and with the established schedules. Observations include verifying that the correct equipment is installed in the specified locations and with proper connections and appropriate workmanship.

For cold-air distribution systems, construction observation should include particular attention to the following:

- Duct insulation is installed with complete coverage and good workmanship, especially at equipment connections and duct hangars.
- Duct insulation vapor retarder covering is installed with complete coverage, with no breaks or punctures.
- Special care is taken to prevent damaging insulation and the vapor barrier.
- Ducts are sealed as specified.
- All insulation is installed before operating duct system with cold supply air.
- Construction dust is removed from the space and from the duct system before operating cooling coils at dehumidifying temperatures.
- Fan-powered mixing boxes are not started up until downstream ductwork is installed.

The construction observations should identify and prevent creative independent interpretation of the design documents on the part of installers. With a thorough and well-documented design, the chances of such freelance interpretation are minimized. If design documentation is not complete, installers will be required to make decisions. In such cases, contractors should be clearly instructed to seek official interpretation from the design engineer, and the engineer must be readily available to provide that guidance.

If a cooperative spirit has been established on the project team, construction observation is seen as an opportunity to clarify requirements rather than an effort to catch mistakes. It should be clear to all parties that it is to everyone's benefit if the design intent is clearly understood and all parts of the system are installed correctly the first time.

Construction observation is typically performed by the design engineer and by the owner's personnel. The design engineer is in the best position to verify that equipment is installed as required by the design. It is important that construction observation services be clearly included in the engineer's scope of work and fee.

The owner's operations and maintenance personnel should also take part in construction observation. Their experience and perspective lead them to notice details that a design engineer might not see. Construction observation also provides operations and maintenance staff with an improved understanding of the system's physical layout, particularly the locations of equipment and devices that may be obscured when construction is complete.

7.5 ACCEPTANCE PHASE

During the acceptance phase of a project, verification tests and functional performance tests are carried out to confirm that all components and subsystems function together to fulfill the design intent. Proper operation of the cold-air distribution system is dependent on the performance of the central cooling plant and fluid distribution systems. Performance tests of distribution systems are typically carried out in conjunction with tests of these other subsystems. This discussion emphasizes testing considerations specifically related to cold-air distribution.

At the completion of the acceptance phase, the system will have been proven and documented to be performing in accordance with the contract documents, and the owner will have complete test results on the system's performance for all operating modes.

7.5.1 VERIFICATION

The verification, or functional operation evaluation process, verifies that all components, equipment, and systems operate in accordance with the contract documents and manufacturers' specifications.

Before verification can begin, all equipment must have been started up and be fully operational. Control systems must be completed, calibrated, and operational. Testing, adjusting, and balancing (TAB) procedures must be complete and TAB reports submitted.

Verification may be applied to every component in the system, from air-handling unit fans to control valves and pressure sensors. Alternatively, a selected sample of components may be tested, with additional tests performed if significant deficiencies are found.

Verification for cold-air distribution systems includes observation of such items as the following:

- Flow rates are as reported in TAB report.
- Fan volume modulates correctly to maintain duct pressure setpoint.
- VAV terminals modulate correctly to maintain room temperature setpoints.
- Control valves modulate correctly to maintain discharge air temperature setpoints.
- Flowmeters and temperature sensors report accurate measurements.

- Pumps, valves, and fans sequence correctly in each operating mode and when changing between modes.

Verification may be performed by the installing contractor or by an independent testing contractor, with observation by the commissioning agent and other members of the project team. Operation and maintenance personnel should also be present to observe verification, thus improving their understanding of components and system function and operation.

7.5.2 FUNCTIONAL PERFORMANCE TESTING

Functional performance testing demonstrates that the cold-air distribution system performs as required by the final statement of design intent and the contract documents. The functional performance test is required to verify that all components of the system function together as designed and that the system meets the cooling load while satisfying the specified demand and energy targets.

Basic questions to be answered by functional performance testing of a cold-air distribution system are:

- Does the system function according to the design intent?
- What is the peak cooling capacity of the system?
- What is the efficiency of the system in meeting the load?

Functional performance testing also helps establish final revised operating setpoints for control parameters, such as supply air temperature setpoints, duct pressure setpoints, ramp-down schedules, minimum outside air parameters, etc.

While the verification tests evaluate the operation of individual components, the functional performance test evaluates all the components functioning together as a system. A cold-air distribution system performance test should evaluate each air-handling unit's ability to meet its peak load in each zone. The performance test will verify:

- Cooling coil performance
- Fan volume control
- Supply air temperature control
- Space air distribution
- Space temperature and humidity control
- Load on the central plant

7.6 TRAINING

Training of operators is essential for cold-air distribution systems. Most operating and maintenance personnel are not familiar with the general concepts or the details of cold-air distribution. If not properly trained, they may operate the systems based on misconceptions of the design intent, which generally results in poor performance. Training should include fundamentals of cold-air distribution and a complete review of the design intent for the particular installation.

Specific maintenance issues to be addressed in training include:

- the critical need to maintain duct insulation and vapor barriers intact,
- maintenance requirements for glycol heat transfer fluids, and
- general logic and specific details of ramp-down and supply temperature reset, and other control sequences.

Involving operators in the construction and acceptance phases of commissioning is highly recommended as a part of training. Participating in testing allows operators to become familiar with thermal storage concepts, as well as the design intent and specific operating characteristics of their system.

7.7 OPERATION

During the system operation and optimization phase, operators gain experience with the system and adapt its operation to achieve optimum performance. Any changes in the system must be recorded in as-built documents. Changes in the use and function of facilities must be monitored and documented, and appropriate adjustments made in system operation. For example, changes in occupant density or activity, occupancy schedules, or other load factors may require modifications in cold-air distribution system operating strategies.

Tenant improvement work in a speculative office building is typically carried out after the core system has been installed. The engineers and contractors for tenant improvement work are not necessarily those involved in designing and installing the core system. Particularly for a central fan system using cold-air distribution, inappropriate distribution designs in tenant spaces can result in the core system being unable to meet the needs of the last portion of the building to be leased.

It is critical that tenant improvement project teams thoroughly understand the requirements and limitations of the core system. This information should be included in the statement of design intent for the core system and must be provided to each tenant improvement team.

7.8 REFERENCES

ASHRAE. 1996. ASHRAE Guideline 1-1996, The HVAC Commissioning Process. Atlanta: American Society of Heating, Refrigerating and Air-Conditioning Engineers, Inc.

CHAPTER EIGHT

ECONOMIC EVALUATION

8.1 INTRODUCTION

The decision to use cold-air distribution technology to provide space conditioning is grounded in economics. The use of cold-air distribution is motivated by the prospect of economic benefit relative to other cooling systems. The benefit might be a reduction in first cost, operating cost, or both, or an improvement in the environment maintained by the system. The economic evaluation attempts to put a relative value on first costs incurred in the present and operating costs incurred over a number of years to provide a meaningful comparison of alternatives.

The inputs to the economic evaluation are the first costs and operating costs of a number of alternatives and the owner's financial analysis criteria. The result is a ranking of the alternatives in terms of economic attractiveness.

This section provides an overview of the economic considerations associated with cold-air distribution design.

8.2 METHODS OF ECONOMIC ANALYSIS

In any comparison of alternative systems, there will be differences in first costs and operating costs that must be balanced against each other to determine which option is most advantageous to the owner. A number of methods of economic analysis are used to make these comparisons. Sophisticated owners will have their own criteria for judging economic alternatives. However, in many cases, the manner in which the economic information is presented can have an effect on which alternative is chosen.

The simple payback method is a commonly used economic analysis technique. This approach simply evaluates the ratio of the annual operating cost savings to the incremental cost of the project. The project's first and operating costs are generally compared to a base case, which is often the standard conventional design.

The simple payback method is a quick and easy way to evaluate a project's benefits. Unfortunately, it does not accurately assess all of a project's economic implications. For example, it does not consider the cost of money for the initial investment and it ignores inflation and the time value of money.

More sophisticated methods of economic analysis, such as life-cycle costing and cash flow analysis, use the principles of present value analysis to account for the time value of money. These methods are recommended for all projects where accurate economic comparisons are required. Economic analysis is discussed in detail in chapter 33 of the *1995 ASHRAE Handbook—Applications* (ASHRAE 1995).

8.3 FIRST COST ESTIMATION

First cost estimates for cold-air distribution systems must consider costs for ductwork, terminal units and diffusers, air-handling units including cooling coils, pumps, and piping. Electrical costs may also differ from conventional systems.

In general, the most accurate estimates are developed by selecting equipment for the cold-air distribution and conventional alternatives and obtaining quotes or budget estimates from contractors or suppliers.

Ductwork costs are highly dependent on the specific layout, including number of straight runs and number and type of fittings. For preliminary estimates, it may be acceptable to use a normal rule of thumb for conventional duct costs and assume a savings of 20% to 30% for cold-air distribution. Note that the cost reduction will generally not be proportional to the reduction in pounds of sheet metal.

The most accurate method is to size the duct system for both conventional temperature air and cold-air distribution and perform a complete takeoff and cost estimate on both systems. If it is impractical to follow this approach for the whole system, a representative portion of the system can provide an indication of the percentage cost savings for the cold-air distribution option.

Reduced insulation costs due to reductions in the total duct surface area typically offset the cost of increased insulation thickness.

Terminal units and diffusers for cold-air distribution may be more or less expensive than for a conventional design. The total volume of air handled, and the size of the equipment, is less. However, there may be additional requirements for unit insulation or throw performance that tend to increase the cost. Cost estimates can most easily be developed by obtaining quotes from suppliers for the equipment options.

Air-handling units for cold air distribution are generally smaller and less expensive, even when the cooling coils are more expensive.

Pumps and piping are generally smaller in cold-air distribution systems than in systems using smaller fluid temperature ranges. Piping is typically one size smaller for a 20°F (11°C) range than for a 12°F (6.7°C) range. Pipe insulation thickness is increased by 0.5 in. (13 mm) for pipes carrying fluid below 40°F (22°C), consistent with ASHRAE Standard 90.1 (ASHRAE 1989).

Electrical costs for cold-air distribution systems may be reduced due to smaller size or a smaller number of air-handling unit fans. Systems using fan-powered mixing boxes may have increased electrical costs.

8.4 OPERATING COSTS

Operating cost estimates should consider the fan energy savings, as well as any cooling energy penalty due to lower temperature operation or the reduced availability of economizer operation. The calculation of estimated operating costs is discussed further in chapter 9.

8.5 NONCOOLING SYSTEM COST CONSIDERATIONS

The use of reduced supply air temperatures can affect building costs not directly related to the mechanical systems. These effects include possible reductions in personnel costs through improved productivity, and reductions in building structure and equipment costs that are related to the space required for ductwork.

Where the lower humidity levels lead to improved occupant comfort, increases in productivity can have benefits that far exceed the cost of energy for space conditioning. For example, the total cost of wages and benefits for employees in a typical office occupancy is generally on the order of \$100/ft^2 (\$1000/m^2), while the energy cost for comfort conditioning is on the order of \$1/ft^2 (\$10/m^2). If improved comfort leads to a 1% increase in employee productivity, the benefit is equivalent to a savings of 100% of the energy cost. Of course, these benefits are difficult to predict or to demonstrate. However, many owners are convinced that the productivity benefits of improved comfort are real and that they justify investment in high-quality space-conditioning systems.

Cold-air distribution technology can lead to reduced building construction costs. If the required space above the finished ceiling is determined by duct clearance constraints, the use of cold-air distribution can reduce that requirement by three inches or more. Particularly in high-rise buildings, the resulting decrease in floor-to-floor height can reduce the cost of the building envelope and structural components, as well as plumbing, electrical, and elevator systems. For a building with 15 floors or more, the height reduction may lead to a 3% to 5% decrease in the total building construction cost (Dorgan et al. 1995).

Even greater savings can be achieved if the smaller duct sizes allow architects and structural designers to use innovative design approaches. For example, downsized ducts may be small enough to be routed through beam penetrations rather than under the beams. This approach allows designers to use taller and thinner structural members. Little clearance is required between ceiling and beams, and a large reduction in overall floor-to-floor height is possible.

The smaller fans used for cold-air distribution may allow architects to reduce the space allocated to mechanical rooms, providing additional usable floor area to the owner.

8.6 DESIGN DECISIONS

Every design parameter selection or equipment-sizing decision has an impact on the system's economic performance. However, the possible choices for many decisions are limited within fairly narrow ranges by the sizes or capabilities of readily available equipment. The primary choices facing the designer are the supply air temperature and the cooling fluid temperature range.

The selection of supply air temperature has the most far-reaching effects on equipment size and energy consumption. Reducing the supply air temperature

- reduces the size and cost of fans and ducts;
- reduces fan energy consumption and cost;
- may increase cooling energy, due to increased dehumidification and decreased economizer availability;
- increases the cost of cooling coils; and
- may increase insulation costs.

In many cases, the selection of supply air temperature is limited by external factors. For example, a maximum space relative humidity of 40% or a maximum duct height of 12 in. (305 mm) is a constraint that limits the possible range of supply temperatures. Other constraints might be the maximum depth of cooling coil for effective maintenance or the minimum available chilled-fluid supply temperature.

The fluid temperature range affects the size and cost of fluid distribution pumps and piping and cooling coils. The achievable range is determined primarily by the performance of available cooling coils. For most cold-air distribution designs, a 20°F (11°C) range is easily achievable and is recommended. In many cases, a 24°F (13°C) range is possible for additional savings in water distribution equipment and energy costs. For designers whose experience is primarily with 10°F to 12°F (5°C to 7°C) ranges and who may prefer not to depart so far from previous practice, a minimum 16°F (9°C) range is recommended.

8.7 REFERENCES

ASHRAE. 1989. *ASHRAE Standard 90.1, Energy efficient design of new buildings except low-rise residential buildings*. Atlanta: American Society of Heating, Refrigerating and Air-Conditioning Engineers, Inc.

ASHRAE. 1995. *1995 ASHRAE handbook—Applications*. Atlanta: American Society of Heating, Refrigerating and Air-Conditioning Engineers, Inc.

Dorgan, C.E., C.B. Dorgan, S.B. Leight. 1995. Cool Storage Total Building Construction Cost Benefits—An Owners and Architect's Guide. EPRI IR - 104521. Palo Alto: Electric Powers Research Institute.

CHAPTER NINE

COMPUTER MODELING

9.1 INTRODUCTION

The primary purposes of computer modeling for cold-air distribution systems are to calculate design loads and to estimate the annual energy consumption of a given system. Many programs perform both design load and annual energy calculations from the same set of inputs.

The use of computer programs for HVAC load and energy calculations is discussed extensively in the following references:

- *1993 ASHRAE Handbook—Fundamentals,* chapter 28, "Energy Estimating Methods"
- *1995 ASHRAE Handbook—HVAC Applications*, chapter 36, "Computer Applications"
- *Air-Conditioning Systems Design Manual*, section 2.5 (Lorsch et al. 1993).

Designers should use care when applying computer simulation programs to optimize cold-air distribution design decisions. Most available programs were developed to model conventional nonstorage cooling systems. These models may provide inaccurate results when applied with extreme supply air temperatures or with storage cooling plants.

The following sections discuss specific considerations for modeling cold-air distribution systems with common simulation programs.

9.2 DESIGN LOAD CALCULATIONS

Designers use computer calculations of design loads to size equipment such as fans, cooling coils, chillers, and other components. When calculating loads for cold-air distribution systems, users should consider the limitations of the program used.

The simplest load calculation programs calculate the space load and outside air ventilation load but do not consider effects of the distribution system on the equipment load. These programs may allow the user to specify distribution system gains, either as a percentage of the total load or as a supply air temperature increase.

More sophisticated programs simulate the distribution system performance. Calculation of fan heat gain is commonly offered, but duct heat gains may be ignored or require user definition. Heat gains from fan-powered mixing boxes may or may not be considered. Most programs do not perform a complete psychrometric analysis as described in section 4.10. The space relative humidity may or may not be a user-definable parameter, and the return air relative humidity generally is not calculated.

The impacts of these limitations on load calculation for cold air distribution vary. If heat gains are not correctly calculated, required supply air volumes may be underestimated, resulting in undersized fans and terminal units. If actual humidity relationships are not taken into account, calculated cooling coil latent loads may be incorrect, resulting in improper coil selections.

Heat gains to supply air are generally provided as input to a program as a fixed temperature gain or as a percentage of the total load. Some programs calculate the fan heat gain separately, while others require that fan heat be included in the general heat gain definition.

Fan-powered mixing boxes add a significant heat gain to the supply air, including fan heat and the additional load imposed by recirculated return air. Designers should include an allowance in the space load for these gains if they are not explicitly calculated by the simulation program.

Designers should take care to specify heat gains for cold-air distribution systems correctly. Fan heat generally causes the same temperature rise as in a conventional system but a higher percentage gain. Duct heat is a roughly equivalent percentage of the total load but a larger temperature rise.

If the load calculation program assumes typical conventional values for space humidity and return air humidity, calculations of cooling coil latent load will not be correct for cold-air distribution systems. Some programs allow the user to specify the space relative humidity. This value should be determined by the methods given in sections 4.3 and 4.10.

As a check on the computer program output, designers should select one zone and carry out a complete psychrometric analysis as discussed in section 4.10. A comparison of the computer output with the results of the psychrometric analysis will show whether the accuracy of the program is acceptable or whether adjustments to the computer output are required.

9.3 ENERGY CALCULATIONS

Computer modeling is often used to compare the annual energy consumption of a number of alternative systems. The alternatives may be different system types, such as an ice storage system with cold air distribution compared to a nonstorage system with conventional supply air temperature. On the other hand, the comparison may be carried out to optimize design parameters such as supply air temperature or to evaluate alternative control strategies. Energy modeling is also used to demonstrate compliance with energy codes.

Modeling considerations for cold-air distribution systems differ according to whether the simulation is comparing different system types or evaluating system design parameters.

9.3.1 COMPARISON OF SYSTEM TYPES

An energy simulation of alternative system types is typically performed for a feasibility study, with the intent of identifying the energy costs of each system. A common comparison is between a nonstorage system with conventional distribution temperatures, an ice storage system with conventional distribution temperatures, and an ice storage system with cold air distribution.

Many commercially available energy analysis programs are capable of providing acceptable accuracy for an initial screening study of the above three options, but obtaining accurate energy consumption estimates for specific system configurations or control strategies may be difficult. For example, many programs offer the option to model cool storage. However, the flexibility of these models is, in general, limited. The programs typically offer one or two storage types from which to choose. Energy calculations are based on certain assumptions about system operation. Designers often find that it is difficult to model the specific system and operating strategy that they have selected.

In general, modeling of cool storage and cold-air distribution systems is not a simple process. Designers who wish to obtain accurate comparisons of system energy use can expect to spend significant time and effort experimenting with the simulation program to learn the assumptions embedded in the program. This experimentation generally consists of running and rerunning models with various combinations of system parameters, and analyzing the results to identify how the program treats changes in each parameter.

9.3.2 OPTIMIZING DESIGN PARAMETERS

Designers may wish to optimize their selection of design parameters or control strategies by simulating various options. A detailed and accurate model is required to evaluate the effects of variations in design parameters, such as supply air temperature, fluid supply temperature, or fluid temperature range, or control strategies, such as supply air reset.

A true optimization would consider all the possible combinations of the design parameters in question. In practice, a designer will generally select a limited number of combinations and compare the energy consumption. Note that a complete optimization analysis will consider the life-cycle cost of each of the options analyzed, as discussed in chapter 8.

9.3.3 MODELS

Most commercially available energy analysis programs are designed using the *fixed schematic* simulation technique (ASHRAE 1995, chapter 36). A program using this approach consists of a calculation procedure that defines a given set of systems. The general system schematic is selected from among the available choices, and equipment performance characteristics are defined by the user. The various programs differ in the number and types of systems that can be modeled and the options available to the user for describing component characteristics.

The *component relation* approach is organized around components rather than systems (ASHRAE 1995, chapter 36). Each of the components is described mathematically and placed in a library for use in constructing the system. The user input includes the definition of the system schematic as well as equipment characteristics and capacities.

If the desired level of detail is not available from commercial programs, designers may choose to construct a custom component relation model using a simultaneous equation-solver program (FCS 1994; MSI 1994). A program of this type allows the user to specify the defining equations for each component of interest, then solves the equations for all components simultaneously. This approach allows a much more accurate simulation of the interactions among components than most fixed schematic-type programs because component descriptions are not limited to predefined options. However, this method requires a thorough understanding of the equations defining component performance and the relationships among the components. Multiple iterations of the simultaneous equation solver, for a range of ambient and load conditions, are required to obtain an annual energy analysis.

9.4 COMPONENT MODELING CONSIDERATIONS

9.4.1 GENERAL CONSIDERATIONS

Energy comparisons of cold-air distribution design options generally involve relatively small changes in overall energy consumption. Accurate models are required to resolve the differences in chiller, pump, and fan energy consumption resulting from changes in design parameters. The primary rule for modeling cold-air distribution systems is:

Any error introduced by inaccuracies in component or system models must be smaller than the differences among the options compared.

For example, the largest contributor to cooling system energy consumption is the chiller. The accuracy of the chiller performance model must be very high because a small percentage error in chiller energy consumption will have a disproportionate effect on the total system consumption.

9.4.2 COOLING PLANT

A key issue in evaluating energy consumption of cold-air distribution systems is the possible increase in chiller energy due to lower fluid supply temperature, increased dehumidification, and reduced economizer availability. To properly assess this issue, the simulation program must accurately model the cooling plant, since the difference in energy consumption between alternative cooling plants is often small.

If the cold-air distribution system is to use a nonstorage cooling plant, modeling is relatively straightforward. The chiller performance curves used to describe the cooling plant must apply to the specific chiller and chilled-fluid temperatures in question. In particular, if the effects of reduced fluid supply temperature are to be evaluated, the temperature dependence of the chiller performance must be accurately represented.

As discussed above, accurate simulation of cool storage plants is difficult. Most programs cannot model the interaction between chiller and storage performance, the effects of chiller-upstream or chiller-downstream configuration, or the performance impacts of chiller-priority, storage-priority, and demand-limiting control strategies.

The details of cool storage system operation may not be important for generalized screening studies. However, for a designer to adequately evaluate alternative design options, the simulation program must be capable of modeling the differences between those options. Engineers wishing to model cold-air distribution systems with cool storage should understand the assumptions behind the cool storage model their program uses.

9.4.3 PUMP ENERGY

Pump energy constitutes a significant fraction of the annual energy consumption of many cooling systems. Differences in pumping energy are important in comparisons involving cold-air distribution systems with high chilled-fluid temperature ranges.

Pump energy is difficult to model in detail with most simulation programs. Typically, pumps may be specified with a constant energy consumption whenever chillers are operating, or with a variable consumption proportional to the percentage of full load. The constant consumption input corresponds to chiller primary pumps and condenser water pumps, which operate at constant volume when chillers are on. The variable-consumption input corresponds to secondary distribution pumps, which operate at variable volume as the load varies. The cubic relationship between power and flow may or may not be considered. Variations in pump operation with the mode of storage operation are also difficult or impossible to model.

Most programs simulate air distribution systems and central cooling plants separately and cannot model the interactions between cooling coil performance and chilled-fluid temperature. Generally, programs implicitly assume that a system's temperature range will be constant over the entire range of part-load operation, although this is rarely true in actual systems.

9.4.4 SUPPLY FAN ENERGY

The modeling of supply fan energy consumption is straightforward and is easily accomplished with most simulation programs. Fan energy comparisons of cold-air distribution options are sensitive to the specified fan total pressure and to the fan performance curves for simulating part-load operation.

For accurate comparisons, designers should specify the appropriate fan total pressure for the systems simulated.

Many programs supply generic part-load performance curves for supply fans, which may or may not be accurate. For example, a commonly used part-load curve for variable-speed drives shows 5% fan power at 20% fan volume, while in fact most drives will not modulate below 30% of full speed. Modelers should confirm with variable-speed drive suppliers that the curves used correctly represent the drive characteristics. Simulations should also consider the minimum duct pressure to be developed by the supply fan.

Modelers should also verify how the simulation treats fan system diversity. Some programs assume that the fan's design flow is equal to the sum of the peak zone flow requirements, which is not true in a system designed for diversity. A model making this assumption will treat a fan designed for 75% diversity and operating at 100% flow as if it were at 75% of full flow. This error causes an underestimate of fan energy and of fan energy savings for cold air distribution.

9.4.5 FAN-POWERED MIXING BOXES

The energy consumption of fan-powered mixing boxes in a cold-air distribution system is typically equal to or greater than that of the supply fans. Therefore, designers should accurately consider this energy when comparing system alternatives.

Fan-powered boxes typically use small fractional-horsepower motors and relatively inefficient fans, leading to low overall efficiencies. Overall fan plus motor efficiency is typically in the range of 5% to 10%.

Simulation programs may require fan power as input data or may calculate fan power from air volume, static pressure, and efficiency. Fan power can be obtained from manufacturers' data or from ARI ratings.

Intermittent mixing box control strategies based on primary air temperature or flow may be difficult to model in some programs.

9.4.6 ECONOMIZER CONTROL

Economizer control sequences compare the outside air temperature or enthalpy with another quantity to determine whether outside air can be used for cooling.

An enthalpy economizer cycle uses outside air for cooling whenever its enthalpy is lower than that of the return air. Other variations compare the outside air dry-bulb temperature with that of the return air or with a fixed changeover temperature.

Modelers should specify an economizer control option that realistically represents the control that will be used by the system simulated. The fixed changeover is the simplest and most commonly used method. The enthalpy economizer maximizes energy savings, but it requires special attention to calibration or it can result in improper control and increased energy consumption.

Programs that do not calculate the actual return air condition, but make an assumption about the space or return air humidity, cannot accurately model an enthalpy economizer with cold air distribution.

9.4.7 SUPPLY AIR RESET CONTROL

A supply air reset sequence resets the supply air temperature to the highest level below some maximum that will satisfy the zone with the highest cooling load. In cold-air distribution systems, this strategy is often used to decrease excessive dehumidification energy, to increase economizer availability, or to maintain a higher level of room air circulation. As loads vary, there may be a relatively large number of hours when the supply air temperature can be reset.

Modeling of supply air reset control is dependent on scheduling of zone loads. In a typical simulation, occupancy schedules are given as general averages for blocks of zones. Lighting and equipment heat gains are commonly described in terms of average watts per square foot. All similar zones will have the same calculated loads for each hour, except where exterior exposures differ.

In a real building, loads vary as people move in and out of zones, turn lights and equipment on and off, and open and close window shades. The actual occupancy and equipment density schedules will show much more variation between zones and over time than the simulation.

Real buildings also show more interaction between exterior and interior zone loads than is typically reflected in simulations. There may be mixing of return air in common plenums, airflow through doorways, or even common open spaces served by both interior and exterior zone distribution.

To accurately simulate the impact of a supply air reset strategy, a model should consider differences between "identical" zones and include some variations in individual zone load schedules. If possible, the model should also account for interactions between interior and exterior zones.

9.5 REFERENCES

ASHRAE. 1993. *1993 ASHRAE handbook—Fundamentals.* Atlanta: American Society of Heating, Refrigerating and Air-Conditioning Engineers, Inc.

ASHRAE. 1995. *1995 ASHRAE handbook—HVAC applications.* Atlanta: American Society of Heating, Refrigerating and Air-Conditioning Engineers, Inc.

FCS. 1994. *Engineering equation solver.* F-Chart Software.

Lorsch, H.G., et al. 1993. *Air-conditioning systems design manual.* Atlanta: American Society of Heating, Refrigerating and Air-Conditioning Engineers, Inc.

MSI. 1994. *Mathcad Plus 5.0.* MathSoft, Inc.

APPENDIX A

SYMBOLS AND CONSTANTS

SYMBOLS

Δp = vapor pressure differential, in. Hg (kPa)
ΔT = temperature difference, °F (°C)
η = efficiency, %
ν = specific volume, ft^3/lb (m^3/kg)
ρ = density, lb/ft^3 (kg/m^3)
A = area, $in.^2$ or ft^2 (mm^2 or m^2)
bhp = fan brake horsepower, hp (kW)
C_s = separation coefficient
D = diameter, in. (mm)
db = dry-bulb temperature, °F (°C)
F = outdoor air fraction required by critical zone.
H = height, in. (mm)
k = insulation thermal conductivity, $Btu \cdot in./h \cdot ft^2 \cdot °F$ ($W/m \cdot K$)
K = velocity decay coefficient
L = length, ft (m)
M = permeance, $gr/h \cdot ft^2 \cdot in.$ Hg ($ng/[s \cdot m^2 \cdot Pa]$)
M_v = mass of water vapor, gr/h (g/s)
NC = noise criteria
p = pressure, in. wg (Pa)
P = perimeter, in. (mm)
P = power, W
q = heat gain, Btu/h (W)
Q = airflow rate, cfm (L/s)
R = required ventilation rate, cfm (L/s)
R = thermal resistance, $°F \cdot ft^2 \cdot h/Btu$ ($m^2 \cdot K/W$)
RH = relative humidity, %
t = thickness, in. (mm)
T = throw distance
T = temperature, °F (°C)
U = overall heat transfer coefficient, $Btu/h \cdot ft^2 \cdot °F$ ($W/m^2 \cdot K$)
V = linear average velocity, fpm (m/s)
vp = vapor pressure, in. Hg (kPa)
W = width of duct, in. (mm)
W = humidity ratio, lb_w/lb_a (kg_w/kg_a)
wb = wet-bulb temperature, °F (°C)
x = distance
X = outdoor air fraction
y = intermediate calculation for duct heat gain
Y = corrected outdoor air fraction

UNIT CONVERSION CONSTANTS

C_1	= 1.1 Btu·min/ft^3·h·°F (1.24 W·s/m^3·°C)
C_2	= 4,840 Btu·min·lb$_a$/h·ft^3·lb$_w$ (3,002 W·s/L)
C_3	= 7,000 gr/lb$_w$ (1 kg/kg)
C_4	= 0.40 Btu·min/in. w.g.·ft^3 (9.99 × 10^{-4} W·s/L·Pa)
C_5	= 2,545 Btu/hp·h (1 kW/kW)
C_6	= 0.364°F/in. w.g. (0.00083°C/Pa)
C_7	= 6,354 in. w.g.·ft^3/hp·min (1,000 Pa·L/s·W)
C_8	= 3.413 Btu/W·h (1 W/W)
C_9	= 12 in./ft (1,000 mm/m)
C_{10}	= 2.4 Btu·ft·min/lb·in.·h (2.01 W·m·s/mm·kg)
C_{11}	= 0.6 Btu·ft·min/lb·in.·h (0.5 W·m·s/mm·kg)
C_{12}	= 144 in.^{2}/ft^2 (0.001 m^2/mm^2) (1000 mm^2·m/L)
C_{13}	= 4.5 lb·min/ft^3·h (1.2 kg/m^3)
C_{14}	= 0.009 lb$_w$·gal/lb$_a$·ft^3 (0.0012 kg$_w$·L/kg$_a$·m^3)

SUBSCRIPTS

a	= air, or ambient air
act	= actual
avg	= average
coa	= outdoor air required bycritical zone
cond	= condensate
cs	= supply air to critical zone
cs	= cross section
dp	= dew point
duct	= duct
e	= entering
ea	= exhaust air
ec	= entering coil
ed	= effective draft
i	= insulation
inf	= infiltration
l	= leaving
lat	= latent
lc	= leaving coil
lf	= leaving fan
ma	= mixed air
mot	= motor
o	= diffuser outlet
oa	= outside air
oa-req	= total outdoor air required
ra	= return air
recirc	= recirculated air
retfan	= return fan

rm	=	room or space
s	=	jet separation
s	=	surface
sa	=	supply air
sens	=	sensible
std	=	standard
tot	=	total
v	=	vapor
w	=	water
x	=	centerline

APPENDIX B

UNITS AND CONVERSIONS

These conversions are for those units most commonly encountered in cold-air distribution design. Additional information and conversions for other units are given in the ASHRAE publication, *SI for HVAC&R* (ASHRAE 1992).

TABLE B-1 CONVERSION CONSTANTS

Dimension	From	Conversion	To
Temperature	°F	(°F-32)/1.8	°C
Temperature Difference	Δ°F	/1.8	ΔK(Δ°C)
Rate of Cooling	ton	× 3.517	kW
	Btu/h	× 0.2931	W
Quantity of Energy	ton·h	× 3.517	kWh
	Btu	× 0.2931	Wh
	Btu	× 1.055	kJ
	Btu/lb	× 2.326	kJ/kg
Chiller Efficiency	kW/ton	3.517/(kW/ton)	COP
Volume Flow	gpm	× 0.0631	L/s
	cfm	× 0.4719	L/s
	cfm	× 0.0004719	m^3/s
Velocity	ft/min	× 0.00508	m/s
	ft/s	× 0.3048	m/s
Pressure	in of water	× 248.8	Pa
Specific Heat	Btu/lb·°F	× 4.184	kJ/(kg·K)
Length or Head Pressure	ft	× 0.3048	m

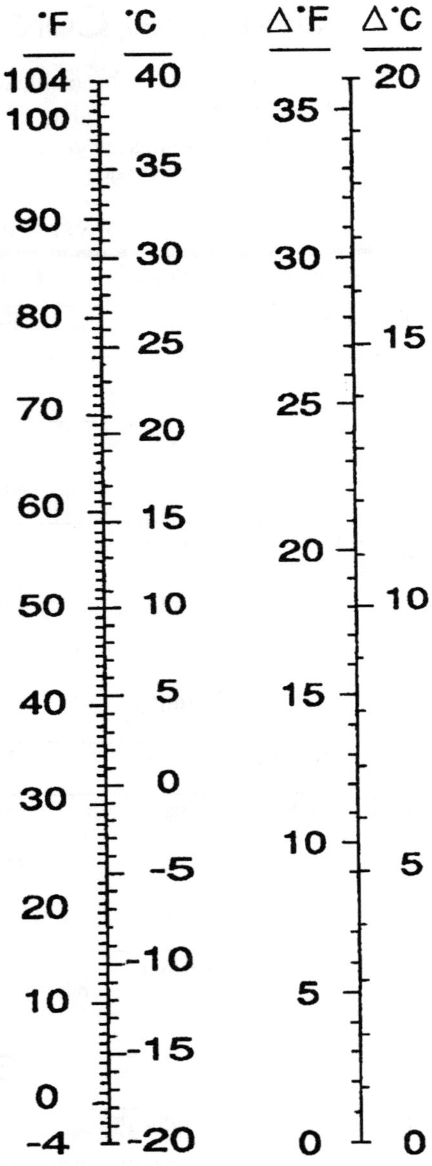

FIGURE B-1 Temperature Conversions.

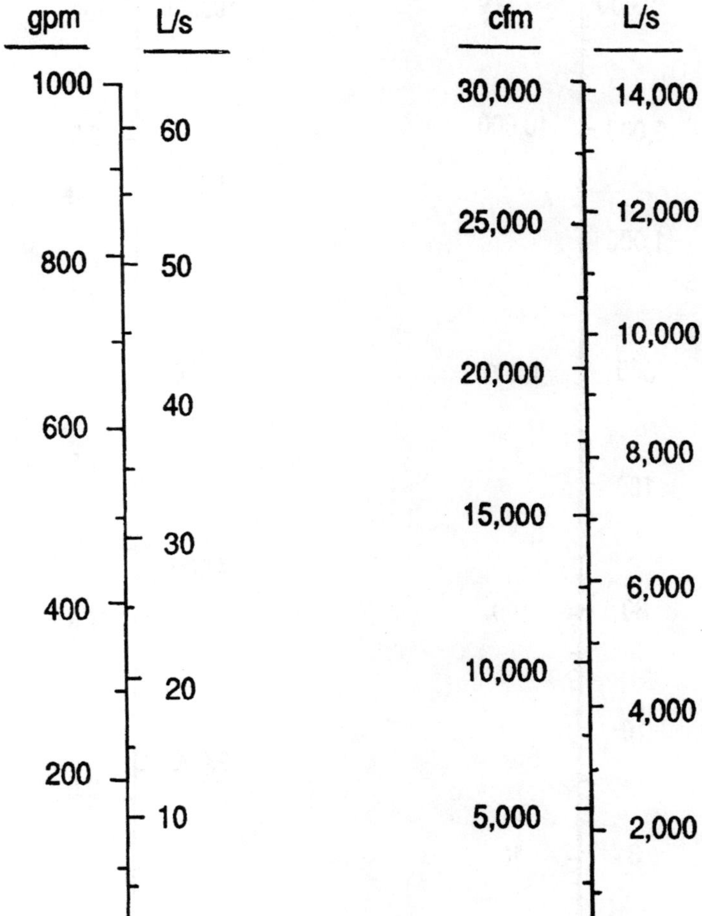

FIGURE B-2 Volumetric flow conversions.

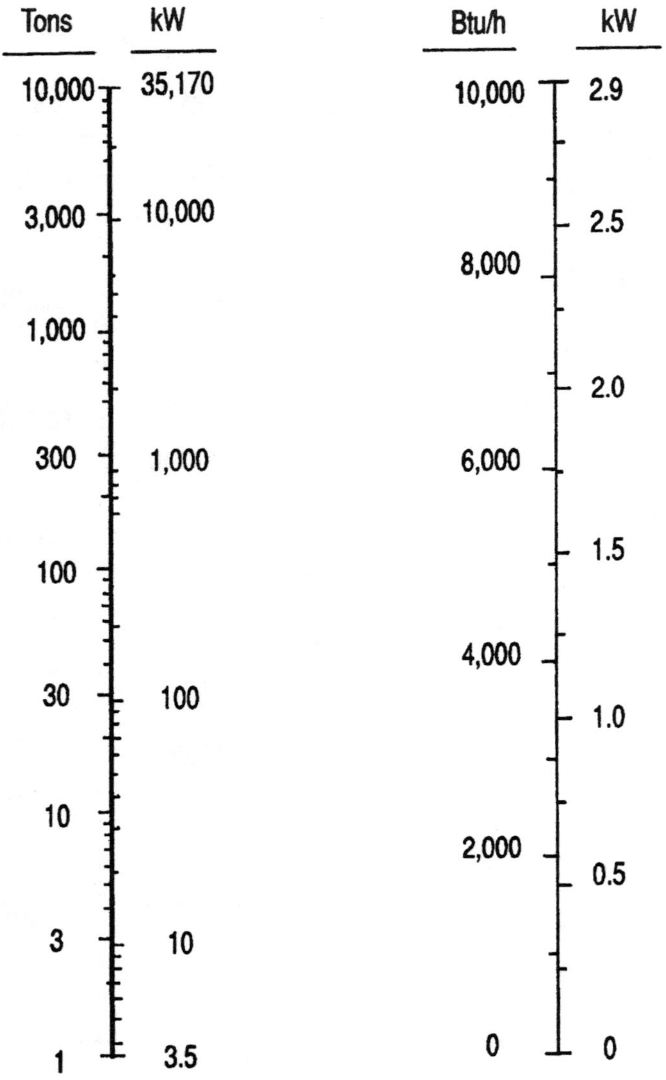

FIGURE B-3 Cooling capacity conversions.

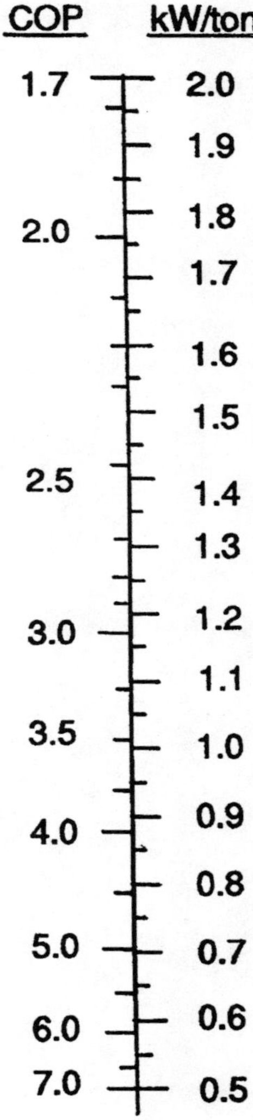

FIGURE B-4 Chiller efficiency conversions.

ANNOTATED BIBLIOGRAPHY

ASHRAE. 1989a. *Guideline 1-1989, Commissioning of HVAC systems*. Atlanta: American Society of Heating, Refrigerating and Air-Conditioning Engineers, Inc.
Provides procedures and methods for documenting and verifying the performance of HVAC systems to ensure that they operate in conformity with the design intent.

ASHRAE. 1989b. *ANSI/ASHRAE Standard 62-1989, Ventilation for acceptable indoor air quality*. Atlanta: American Society of Heating, Refrigerating and Air-Conditioning Engineers, Inc.
Provides procedures and methods for calculating and documenting ventilation air requirements for a conditioned space.

ASHRAE. 1993b. Cold air distribution and thermal storage. *Technical Data Bulletin* 9(2).
A collection of papers presented at the January 1993 winter ASHRAE meeting. Contents include:

- Ethylene Glycol: Its Use in Thermal Storage and Its Impact on the Environment.
- Theoretical Investigation and Practical Applications of Nonisothermic Jets for the Purpose of Ventilating Rooms.
- Total Environmental Quality.
- Cold Air Should be Jet Mixed, Not Distributed.
- Cold Air Distribution, Application, and Field Evaluation of a Nozzle-Type Diffuser.
- Energy Use of Fan-Powered Mixing Boxes with Cold Air Distribution.
- Indoor Thermal Environment of Cold Air Distribution Systems.
- Use of Computer Modeling Techniques to Investigate Aspects Involved in Cold Air Distribution Systems.

Bauman, F.S., T. Borgers, P. LaBerge, and A.J. Gadgil. 1992. *Cold air distribution in office buildings: Technology assessment for California*. Berkeley: Center for Environmental Design Research, University of California.
Report on a study to identify research needs and to assess the current state of practice, and energy and operating cost implications of cold air distribution in California. Includes whole-building energy simulations to compare energy performance of systems using "conventional" and cold air distribution, with and without ice storage, for three California climates. The merits of economizers and fan-powered mixing boxes are also examined. This paper also includes results of an industry survey to assess the perceived strengths and limitations of cold air distribution, perceived barriers to its use, and user experience.

Bauman, F.S., T. Borgers, P. LaBerge, and A.J. Gadgil. 1993. Cold air distribution in office buildings, a technology assessment for California. *ASHRAE Transactions* 99(2): 109-124.
A condensed version of the full report, Bauman et al. 1992.

Bihler, L.A. 1986. Effective methods of building commissioning documentation. *ASHRAE Transactions* 92(2B): 476-482.
Analyzes the nature of building commissioning, the necessity of thorough documentation of this process in order to achieve optimum building system performance, and examples of documentation methods.

Brady, T.W. 1994. Achieving energy conservation with ice-based thermal storage. ASHRAE Transactions 100(1): 1735 - 1745
Shows that ice storage systems can be designed to use less total energy than conventional systems. Achieving high efficiency involves controls, equipment selection, and operating sequences.

Carroll, W.L., and R.J. Hitchcock. 1993. Tuning simulated building descriptions to match actual utility data. *ASHRAE Transactions* 99(2): 928-934.
Describes the details of a method for systematically adjusting the parameters of a simulatable building description in order to match the simulated building energy performance to the consumption of the corresponding actual building.

Catanese, D.L. 1991. An energy analysis of low-temperature air distribution systems and reduced economizer-cycle cooling. *ASHRAE Transactions* 97(1): 843-853.
Presents an analysis of cooling energy differences between systems using 55°F (13°C) and 45°F (7°C) supply air for seven locations. The results show that even when savings from pumps and higher dry-bulb setpoint are neglected, cold-air distribution uses less energy than traditional temperature air distribution in all locations.

Cohen, T. 1994. Providing constant ventilation in variable air volume systems. *ASHRAE Journal* 36(5): 38-40.
Discusses the need to adjust outdoor air dampers as total supply airflow is reduced. Suggests controlling the return air damper to maintain constant mixed-air static pressure as a means to provide constant ventilation airflow.

Craig, N.C., B.W. Jones, and D.L. Fenton. 1993. Glycol concentration requirements for freeze burst protection. ASHRAE *Transactions* 99(2): 200-209.
Results of an experimental study to determine the concentrations of ethylene glycol and propylene glycol required to provide freeze burst protection for heat exchange equipment. Provides

curves showing pressure buildup in tubes for various glycol concentrations versus temperature.

Dorgan, C.E., and C.B. Dorgan. 1995. *Cold air distribution design guide*. EPRI TR-105604. Palo Alto, Calif.: Electric Power Research Institute.
Updated and expanded second edition of Dorgan and Elleson (1988b), including concepts, design considerations, analysis, case studies, and economics.

Dorgan, C.E., C.B. Dorgan, S.B. Leight. 1995. Cool Storage Total Building Construction Cost Benefits—An Owners and Architect's Guide. EPRI IR - 104521, Palo Alto: Electric Powers Research Institute.
Evaluates the construction cost savings related to ice storage and cold air distribution, including those not directly associated with the mechanical systems.

Dorgan, C.E., and J.S. Elleson. 1987a. *Field evaluation of cold air distribution systems*. EPRI EM-5447. Palo Alto, Calif.: Electric Power Research Institute.
Report on short-term monitoring of three early cold air distribution installations. Provides detailed system descriptions and operating data. Includes design recommendations.

Dorgan, C.E., and J.S. Elleson. 1988a. Cold air distribution. *ASHRAE Transactions* 94(1): 2008-2025.
Summary of cold air distribution benefits and design considerations. Includes a summary of the results given in Dorgan and Elleson (1987a).

Dorgan, C.E., and J.S. Elleson. 1988b. *Cold air distribution design guide,* 1st ed. EM-5730. Palo Alto, Calif.: Electric Power Research Institute.
Detailed presentation of cold air distribution design considerations, including design concepts, equipment selection, analysis, and economics.

Dorgan, C.E., and J.S. Elleson. 1989. Design of cold air distribution systems with ice storage. *ASHRAE Transactions* 95(1): 1317-1322.
Summary of design guidelines presented in Dorgan and Elleson (1988b).

Dorgan, C.E., and J.S. Elleson. 1993. *Design guide for cool thermal storage*. Atlanta: American Society of Heating, Refrigerating and Air-Conditioning Engineers, Inc.
Comprehensive design manual for evaluation and design of cool storage systems. Includes cool storage fundamentals, detailed discussion and comparison of cool storage technologies, economic analysis, design examples.

Dorgan, C.E., J.S. Elleson, and M.S. Downey. 1990. *Detailed field evaluation of a cold air distribution system.* EPRI CU-6690. Palo Alto, Calif.: Electric Power Research Institute.

Report on field monitoring of the cold air distribution system on a typical floor of a 12-story office building in southern California. Energy consumption of fan-powered mixing boxes was found to offset the supply fan energy savings resulting from the reduced supply air temperature. Includes a system description, measured results, and calculated energy consumption for several alternative approaches to supply air mixing and distribution. Provides recommendations for improved operation and control. Volume 2 contains complete performance data in graphical form.

Elleson, J.S. 1991. High-quality air conditioning with cold air distribution. *ASHRAE Transactions* 97(1): 839-842.

Summary of cold air distribution benefits, comparison of space air distribution options, advantages of direct supply of cold primary air to the space.

Elleson, J.S. 1993. Energy use of fan-powered mixing boxes with cold air distribution. *ASHRAE Transactions* 99(1): 1349-1358.

This paper shows that if continuously operating fan-powered mixing boxes are used in a system to accommodate cold air distribution, their energy consumption can offset the savings resulting from downsized central fans. Includes results of field monitoring, evaluation of equipment ratings, and theoretical calculations. Provides options for minimizing fan energy in cold air distribution systems.

Engel, J.A., and A.T. Kirkpatrick. 1993. Experimental determination of the airflow performance of a variable-area radial diffuser. *ASHRAE Transactions* 99(2): 759-769.

This paper reports on an experimental study of the airflow performance of a variable-area radial diffuser. The diffuser's performance was characterized by determination of the velocity decay coefficient, *K*, and the virtual origin, x_p, for various flow rates and outlet areas.

Fields, W.G. 1989. Operation and cost advantages of a liquid overfeed low-temperature air distribution system. *ASHRAE Transactions* 95(1): 1300-1307.

Discusses liquid overfeed refrigeration and its application with cold air distribution.

Fields, W.G., and D.E. Knebel. 1991. Cost effective thermal energy storage. *Heating/Piping/Air Conditioning*, July, pp. 59-72.

Includes guidelines for load calculations, use of induction diffusers for direct supply of cold primary air to the space, design recommendations for cold air distribution systems. Discusses indoor air quality and condensation concerns. Includes descriptions and cost analyses for several cold air distribution projects.

Filardo, M.J. 1991. Proper ventilation of offices and conference rooms. *ASHRAE Journal* 33(9): 10-14.

Illustrates the application of the equation of multiple spaces from ASHRAE Standard 62-1989, which allows calculation of the required outdoor air fraction for a system that serves zones with different occupant densities.

Filardo, M.J. 1993. Outdoor air: How much is enough? *ASHRAE Journal* 35(1): 34-38.

Describes a method for maintaining the required minimum ventilation air quantity in a VAV system serving multiple spaces with different occupant densities under variable load conditions.

Gatley, D.P. 1992. *Cool storage ethylene glycol design guide.* EPRI TR-100945. Palo Alto, Calif.: Electric Power Research Institute.

Provides engineering data and details for use in designing, operating, and maintaining cool storage systems using 25% by weight ethylene glycol.

Harmon, J.J., and H.C. Yu. 1989. Design considerations for low-temperature air distribution systems. *ASHRAE Transactions* 95(1): 1295-1299.

Provides design recommendations for cold air distribution systems.

Harmon, J.J., and H. C. Yu. 1993. Cold air distribution and concerns about condensation. *ASHRAE Journal* 35(5): 34, 36, 38, 40.

Addresses three concerns for condensation: migration, condensation inside building, and condensation in equipment rooms. Provides an analysis of insulation requirements and recommendations.

Hassani, V., T. Malmstrom, and A.T. Kirkpatrick. 1993. Indoor thermal environment of cold air distribution systems. *ASHRAE Transactions* 99(1): 1359-1385.

The behavior of cold air jets, specifically jet momentum and jet separation, are reviewed for application to cold air distribution systems. A simple model of cold air jet separation is developed and combined with jet throw equations.

Hittle, D.C., and A. Bhansali. 1990a. *Expected energy use of ice storage and cold air distribution systems in large commercial buildings.* Palo Alto, Calif.: Electric Power Research Institute.

Analysis of energy use and operating cost of several system options, in various climates under a number of utility role schedules.

Hittle, D.C., and A. Bhansali. 1990b. Estimated energy consumption and operating cost for ice storage systems with cold air distribution. *ASHRAE Transactions* 96(1): 418-427.
Summary of results presented in Hittle and Bhansali (1990a).

Hittle, D.C., and R.W. Haines. 1993. *Control systems for heating, ventilating, and air conditioning,* 5th ed. New York: Van Nostrand Reinhold.
Textbook on HVAC controls. Covers basic control theory, control devices, control systems, digital control, retrofits of existing control systems, and dynamic response and cooling.

Hittle, D.C., and T.R. Smith. 1994. Control strategies and energy consumption for ice storage systems using heat recovery and cold air distribution. *ASHRAE Transactions* 100(1): 1221-1229.
Report on a simulation study investigating the energy consumption of an ice storage system, with and without heat recovery, and with and without cold air distribution.

Int-Hout, D. 1992. Low temperature air, thermal comfort and indoor air quality. *ASHRAE Journal* 34(5):34-35, 38-39.
Discusses impact of reduced humidity on comfort, diffuser sizing issues, room air motion. Provides analysis of ventilation requirements. Discusses effect of air density on errors in balancing measurements.

Int-Hout, D. 1993. Total environmental quality. *ASHRAE Transactions* 99(1): 960-967.
Discusses evaluation of, and relationships among, the environmental factors of thermal comfort, acoustics, indoor air quality, and air distribution.

ITSAC. 1988a. Field evaluation of cold air distribution. *ITSAC Technical Bulletin,* E.1000. International Thermal Storage Advisory Council, February.
Summary of Dorgan and Elleson (1987).

ITSAC. 1988b. Cold air distribution design. *ITSAC Technical Bulletin,* E.1001. International Thermal Storage Advisory Council, April.
Summary of *EPRI Cold Air Distribution Design Guide* (Dorgan and Elleson 1988b).

ITSAC. 1990a. Low temperature air projects in England. *ITSAC Technical Bulletin,* E.1002. International Thermal Storage Advisory Council, August.

ITSAC. 1990b. Expected energy use of ice storage systems in large commercial buildings. *ITSAC Technical Bulletin,* E.1003. International Thermal Storage Advisory Council, November.
Summary of Hittle and Bhansali (1990a).

ITSAC. 1991. Detailed field evaluation of a cold air distribution system. *ITSAC Technical Bulletin*, E.1004. International Thermal Storage Advisory Council, January.

Summary of Dorgan et al. (1990).

ITSAC. 1992. Cost impacts of low temperature air distribution with ice storage. *ITSAC Technical Bulletin,* E.1005. International Thermal Storage Advisory Council, May.

ITSAC. 1994. Considerations regarding the operational characteristics of cold air diffusers. *ITSAC Technical Bulletin*, E.1006. International Thermal Storage Advisory Council, January.

Summary of Kirkpatrick and Hassani (1992).

Ke, Y., and S. Mumma. 1996. A generalized multiple-spaces equation to accommodate any mix of close off and fan-powered VAV boxes. *ASHRAE Transactions* 102(1).

Presents a generalized form of the multiple-spaces equation of ASHRAE Standard 62-1989, that applies to systems using fan-powered mixing boxes to recirculate return air.

Kirkpatrick, A.T., ed. 1994. *Proc. EPRI Workshop—Advances in cold air distribution technology.* Palo Alto, Calif.: Electric Power Research Institute.

A collection of papers presenting the results of the latest research and successful experiences in cold-air distribution technology. The proceedings contain the following:

- Recent research and development results in cold air distribution, including airflow visualization, and cold-air diffuser performance.
- Information about cold-air distribution hardware.
- Case studies detailing building performance with cold-air distribution systems.
- Review of the commercialization and status of cold-air distribution technologies.

Kirkpatrick, A., and V. Hassani. 1992. *Determination of the operational characteristics of cold air diffusers.* EPRI TR-101480. Palo Alto, Calif.: Electric Power Research Institute.

Results of a study of cold air wall jets, intended to provide information to improve the design and selection of diffusers for cold air distribution. Includes:

- Infrared imaging technique, and two- and three-dimensional computer graphic techniques, for visualizing airflow and temperature patterns in a room.
- Experimental and computational determination of velocity and temperature decay and separation point of buoyant wall jets.

- Simplified algorithm for calculating the separation point for various diffusers.
- Development of a similar jet theory to guide the application of cold-air distribution technology using existing throw models.
- Extension of existing air diffusion performance index (ADPI) data to cold-air systems.
- New air change effectiveness model for the design and analysis of ventilation systems.
- Design guideline for cold-air diffuser selection.

Knebel, D.E. 1994. Optimum supply air temperature with conventional and ice thermal storage. *Advances in cold air distribution*. Fort Collins: Colorado State University.
Illustrates a method for determining the supply air temperature that minimizes total system energy consumption for a given configuration of chiller, fans, and pumps at a given load condition.

Knebel, D.E., and D.A. John. 1993. Cold air distribution, application, and field evaluation of a nozzle-type diffuser. *ASHRAE Transactions* 99(1): 1337-1348.
Provides application guidelines for preventing condensation and ensuring room comfort. Discusses application and field testing of a high-induction diffuser, and provides throw and drop data. Velocity, temperature, effective draft temperature, and ADPI data are presented.

Landry, C.M., and C.D. Noble. 1991a. Case study of cost effective low-temperature air distribution and ice thermal storage. *ASHRAE Transactions* 97(1): 845-859.
System description and cost comparison with a "conventional" temperature, hydronic heat pump system.

Landry, C.M., and C.D. Noble. 1991b. Making ice thermal storage first-cost competitive. *ASHRAE Journal* 33(5): 19-22.
Condensed version of Landry and Noble (1991a).

Lorsch, H.G., et al. 1993. *Air-conditioning systems design manual.* Atlanta: American Society of Heating, Refrigerating and Air-Conditioning Engineers, Inc.
Provides practical fundamentals of the design of air-conditioning systems. Intended to bridge the gap between actual design and the theoretical knowledge acquired from college courses and generally available reference books.

MacCracken, C.D. 1993. Cold air should be jet mixed, not distributed. ASHRAE Transactions 99(1): 1333-1336.
Advocates the use of small-diameter, vertical-discharge, high-velocity diffusers on walls to achieve better entrainment and room air mixing.

MacCracken, C.D. 1994. Cold air systems: Sleeping giant. *Heating/Piping/Air Conditioning,* April, pp. 59-62.

General information about cold air distribution and some of the common myths about it, with some information on high-velocity jets.

McQuiston, F.C., and J.D. Spitler. 1992. *Cooling and heating load calculation manual,* 2d ed. Atlanta: American Society of Heating, Refrigerating and Air-Conditioning Engineers, Inc.

Comprehensive manual including theory, procedures, and data for calculation of cooling and heating loads for nonresidential buildings.

Meckler, G., and W. Janus. 1993. The benefits of low temperature HVAC. *Consulting-Specifying Engineer*, November.

This article compares a traditional and a cold-air delivery system, examining both indoor air quality and system efficiency.

Miller, P. 1991. Diffuser selection for cold air distribution. ASHRAE *Journal*, September, pp. 32-36.

Review of the ADPI method for selecting diffusers and reevaluation of original ADPI research data for cold-air distribution application.

Mirth, D.R., S. Ramadhyani, and D.C. Hittle. 1993. Thermal performance of chilled-water cooling coils operating at low water velocities. *ASHRAE Transactions* 99(1): 43-53.

Experimental results show that chilled-water cooling coils operating at low water-side Reynolds numbers do not perform as well as predicted by manufacturers' software. An alternative heat transfer correlation is presented that minimizes this problem.

Poz, M.Y. 1993. Theoretical investigation and practical applications of non-isothermic jets for the purpose of ventilating rooms. *ASHRAE Transactions* 99(1).

Discussion of calculation of ventilating jet behavior, the use of jets at -25°C to 5°C for ventilating rooms, and methods of calculating the parameters of the thermal microclimate within rooms, primarily velocity and temperature of jets.

Rose, P.M., and M.J. Seymour. 1993. Use of computer modeling techniques to investigate aspects involved in cold air distribution systems. *ASHRAE Transactions* 99(1): 1366-1374.

Describes the use of a computational fluid dynamics model to determine temperature and velocity conditions in an interior space.

Scofield, M., and G. Fields. 1989. Joining VAV and direct refrigeration. *Heating/Piping/Air Conditioning,* September, pp. 137-152.

Describes the use of liquid overfeed refrigeration in a cold-air distribution system.

SMACNA. 1985. *HVAC air duct leakage test manual,* 1st ed. Vienna, Va.: Sheet Metal and Air Conditioning Contractors' National Association, Inc.
Provides standard methods for rating and measuring leakage from air ducts.

Tackett, R.K. 1989. Case study: Office building uses ice storage, heat recovery, and cold air distribution. *ASHRAE Transactions* 95(1): 1113-1121.
Detailed system description, including design parameters and control sequences.

Tamblyn, R.T. 1992. Common-sense cooling coils. *Heating/Piping/Air Conditioning,* December, pp. 49-57.
Detailed analysis of the effect of chilled-water range on system first cost and operating cost. Concludes that the optimum range is often much higher than the typical 12°F (–11.1°C).

Tavakkol, S., M.H. Hosni, P.L. Miller, and H.E. Straub. 1994. A study of isothermal throw of air jets with various room sizes and outlet configurations. *ASHRAE Transactions* 100(1).
Experimental testing of throw and comparison of results with existing correlations.

Zhang, J.S., R. Zhang, Z. Li, C.Y. Shaw, L.L. Christianson, and L.H. Sparks. 1994. An experimental study of the ventilation performance of cold air distribution systems. *ASHRAE Transactions* 100(2).
Three types of diffusers were tested with 38°F (3.3°C) and 55°F (12.7°C) supply air in a full-scale test room, with and without internal furnishings. Local mean age of air, air diffusion effectiveness, and contaminant-removal efficiency were measured. Air velocities and temperatures were also measured to calculate the air diffusion performance index (ADPI). The cold-air distribution system provided satisfactory air quality and thermal comfort conditions to the occupied region of the room, and no appreciable differences were found in these conditions between the cold-air and "conventional" systems.